RECENT IMPROVEMENTS OF POWER PLANTS MANAGEMENT AND TECHNOLOGY

Edited by **Aleksandar B. Nikolic**
and **Zarko S. Janda**

Recent Improvements of Power Plants Management and Technology
http://dx.doi.org/10.5772/65534
Edited by Aleksandar B. Nikolic and Zarko S. Janda

Contributors

Bernard Tonderayi Mangara, Adela Bara, Simona Vasilica Oprea, Emilija Kisic, Vera Petrovic, Željko Đurović, Alvin Stern, Juraj Králik, Bosko Josimovic, Sasa Milijic, Ming Ding, Jie Huang, Aleksandar B Nikolic

Published by InTech
Janeza Trdine 9, 51000 Rijeka, Croatia

Publishing Process Manager Mirena Calmic
Technical Editor SPi Global
Cover InTech Design team

First published July, 2017
Printed in Croatia
Legal deposit, Croatia: National and University Library in Zagreb

Additional hard copies can be obtained from orders@intechopen.com

Recent Improvements of Power Plants Management and Technology, Edited by Aleksandar B. Nikolic and Zarko S. Janda
p. cm.
Print ISBN 978-953-51-3357-5
Online ISBN 978-953-51-3358-2

For EU product safety concerns:
IN TECH d.o.o., Prolaz Marije Krucifikse Kozulić 3, 51000 Rijeka, Croatia,
info@intechopen.com or visit our website at intechopen.com.

Meet the editors

Editor, Dr Aleksandar B. Nikolic received his BSc, MSc, and PhD degrees in Electrical Engineering in 1991, 1999, and 2009, respectively, from the School of Electrical Engineering, University of Belgrade, Serbia. He worked 10 years in industry, and since 2005, he is with Electrical Engineering Institute "Nikola Tesla," Belgrade, Serbia, as a principal technical associate. In 2013, he was elected for a position of Institute's Scientific Council President. His special fields of interest include power quality and energy efficiency, control of induction motor drives, and industrial automation. He has published over 120 papers and is a reviewer of several international journals. Presently, he is an IEEE senior member, a member of Board of the Serbian Power Electronics Society, and the president of CIGRE-Serbia committee B4.

Co-Editor, Zarko S. Janda was born in 1960. He received his Dipl. Ing., MSc, and PhD (power electronics) degrees in 1984, 1989, and 2004, respectively, from the School of Electrical Engineering, University of Belgrade, Yugoslavia, all in Electrical Engineering. He has been with the EE Institute "Nikola Tesla," Belgrade, since 1984. He was an associate fellow at School of Electrical Engineering, University of Belgrade, Serbia, between 1993 and 2004. He was a professor at the High School of Professional Studies in Electrical Engineering and Computer Science, Belgrade, Serbia, between 2003 and 2010. Presently, he is a chairman of Serbian IEEE PES Chapter, chairperson of CIGRE-Serbia committee B4, and member of Board of the Serbian Power Electronics Society.

Contents

Preface

Global processes of liberalization and deregulation of energy sector have established new technical and technological requirements to the research and development centers all over the world. Imperative requirements include increase of energy efficiency, reliability, and availability of energy resources. Requirements of the modern energy market are integrations of several scientific disciplines and technologies. This book tends to give some aspects of recent improvements of power plant management and technology, covering several approaches taken from different technologies and disciplines. It starts from maintenance and reliability improvement of thermal power plants, through safety as a must for nuclear power plants, to the environmental impact of various power plants independently on their size and the type of fuel used including renewables. At the end, it is shown how renewable sources could be used in production plant for clean energy fuel.

The book is organized in the following sections. After introduction, the first section covers plant management, maintenance, and reliability—three main requirements for modern and efficient power-producing plants. All three chapters in this section are devoted to thermal power plants, still dominant in most power systems in the world. The first chapter in this section underlines the importance of the key performance indicator (KPI) computation for power plant management in order to continuously monitor and improve the business and technological processes. Today, companies especially from energy sector prove that they have reduced maintenance costs by 30% by moving from the concept of corrective maintenance to the predictive maintenance. Innovative predictive maintenance technique is described in the second chapter of this section. The proposed technique is tested on a specific problem that occurs when time-based maintenance is applied on grinding tables of the coal mill leading to the great material losses because of frequent shutdowns of the entire coal grinding subsystem, as well as the possibility that the failure occurs before replacement. Results of this technique could further yield to establishment of optimal maintenance policy in thermoelectric power plant. Reliability is one of the most important requirements of today's power plants. For complex systems, such as steam power plants, it is required to translate system reliability requirements into detailed specifications for all components that constitute the system. In order to develop a model for estimating the reliability index and evaluating the availability index for a coal-fired generating power station, in the last chapter of this section, the graph theoretical analysis and a graph's characteristic polynomial are presented and illustrated by examples. The second section talks about nuclear power plants (NPP), where safety and improvement of technology are up-to-day topic in the field. The last accidents of the NPP in Chernobyl and Fukushima, besides the reason that caused the accident, raise attention of the researches to verify the safety level of the NPP structures. Further improvements in thorium/uranium fuel usage in high-temperature gas-cooled reactors are explained in the next chapter in this section, where, as a result of the rapid development of nuclear energy, thorium has been analyzed because of its abundant reserves and excellent physical properties.

Environmental aspects of power plants are deeply analyzed in the third section. The chapter focuses on the consideration of aspects of environmental impact of all kinds of power plants, without taking into account the details regarding other aspects of energy sector development. This chapter examines the multi-criteria evaluation (MCE) method for carrying out a strategic environmental assessment (SEA) for power plants, covering the analysis and assessment of the energy sector spatial impacts on the environment and both social and economic elements of sustainable development. The last section deals with the modern world requirements for sustainable means of producing clean energy economically, on a very large scale. This chapter presents a detailed design study of a novel, scalable, self-contained solar-powered electrolytic sodium (Na) metal production plant meant to enable a hydrogen (H_2) fuel, sustainable, closed clean energy cycle. Presented analysis shows that power from renewable source (solar plant) is technically and economically viable for meeting the hydrogen fuel clean energy needs of the large number of motor vehicles in the USA.

We expect that readers from various fields of engineering could find interesting practical solutions for improving power plant's management. This book is useful as a reference especially for industrial engineers involved in plant's maintenance, graduate students in advanced study, and the researches who are working on new technologies toward improvement of power plants' reliability, safety, and overall energy efficiency.

Dr Aleksandar B. Nikolic
Associate Research Professor
Electrical Engineering Institute Nikola Tesla, Belgrade
University of Belgrade, Serbia

Dr Zarko S. Janda
Associate Research Professor
Electrical Engineering Institute Nikola Tesla, Belgrade
University of Belgrade, Serbia

Section 1

Introduction

Chapter 1

Introductory Chapter: Review of Current Research Trends in the Field of Power Plants

Aleksandar B. Nikolic and Zarko S. Janda

Additional information is available at the end of the chapter

http://dx.doi.org/10.5772/intechopen.69980

1. Introduction

Since first AC current high-power hydropower plant was put in operation, built by Nikola Tesla and George Westinghouse in 1895 on Niagara Falls, electrification of the world is dramatically changed. The growing power demand and energy consumption in the last decades require fundamental changes in the process, power production and services. These requirements tend to use both conventional and nonconventional energy generation in order to have power plants useful both economically and environmental friendly to the society. Although new trends in this field focus on producing clean energy from renewable sources, the world's most used fuel in power plants is still coal with 41% of produced global electricity [1]. Coal, oil, nuclear and gas power plants are still dominant for supplying base load in all power grids. Also, energy consumed at power plants for generating electricity is still high. Based on OECD data [2], the amount of electricity supplied to the final consumers was 33% of the total energy consumed at power plants.

In Europe, the largest share of budget spent on research, development and demonstration (RD&D) on energy technology was in energy efficiency and renewable sources [3]. On the other side, in Japan, 39% share of total energy RD&D in 2015 remains in the field of nuclear energy [3]. Regarding nuclear power plant (NPP), more attention is spent on improving safety, especially after accident in Fukushima NPP in March 2011.

2. Energy efficiency and reliability

Improving energy efficiency and reliability goes in several ways. Some of the solutions are to continuously monitor and supervise vital equipment in power plants, like generator transformers, in order to improve maintenance and reduce costs. Additional advantage is decision support, where results taken from online monitoring systems are analyzed by external experts that help plant staff and management to make decision about plant operation when some of the possible malfunction of transformers is detected or expected [4]. This also could yield to proper time schedule of transformer replacement [5].

Modern control systems in power plants cannot be realized without the modern system of monitoring of process parameters or parameters of machines and systems. Continuous monitoring includes continuous monitoring of machine operation (online), automatic storage of information and the possibility of automatic or subsequent processing and analysis. It also includes the generation of specific alarms and their submission to the operator and control system, according to a certain procedure [6]. Diagnostics of the generator are based on a wide range of data from off-line and online testing generators and data analysis. All test data, operating data and data of the machine are stored in a database for generators. Thus, all test data from any laboratory, repairs, unexpected events and failures are available for analysis. The data in the database with each successive inspection and testing are updated. The database is very flexible and has the ability to expand for all possible new types of tests, acquisition of photo records of visual inspections and so on [7].

3. Operation improvement and stability

In virtually all coal preparation operations, mill systems are a critical part of the process to provide economical, reliable and energy-efficient grinding. Operating mills at a slightly lower speed or even a slightly higher speed than line frequency give process engineers the advantage of the mills being optimized for the grade of material and desired throughput of the final process [8]. To get the target boiler power increase in order for 5–10% of rated power, it is necessary to increase the fuel intake and one of the possibilities for that is the coal grinding mill capacity increase [9]. Proposed solution in Ref. [9] is based on enhanced motor voltage supply by increasing frequency, what is possible by medium voltage (MV) inverter. The main goal is to supply motor with rated voltage and frequency in range between 50 and 55 Hz to obtain increase of plant power for 10% by increasing grinding mill capacity. Additional benefits are reduced mechanical stress during start-up and the additional possibility of mill slow running for inspection purposes.

In order to improve power plant stability while operating close to its capability limits, as a requirement of a deregulated electricity market, one solution could be to optimally coordinate the synchronous generators' reactive power outputs in order to maintain the total reactive power delivered by a steam power plant (SPP) or the voltage at a steam power plant high voltage (HV) busbar [10]. In such way, it is possible to aggregate the multimachine power plant into single virtual generator, thus enabling more sophisticated zonal voltage control across power transmission network.

4. Environmental impacts

Environmental impacts of power plants are mainly reflected in emissions of pollutants and greenhouse gases from fossil fuel-based electricity generation. For instance, electricity generation is the fourth highest combined source of NO_x, carbon monoxide and particulate mvatter in the United States [11]. The combustion of coal for power generation produces fly ash, which must be collected prior to discharge to the atmosphere. Electrostatic precipitators are devices used for collecting of fly ash from smoke gases in power plants that use coil as a combustion fuel. The precipitator collection efficiency can be expected to exceed 99.5%. Most existing electrical precipitators are developed with classical continual power supply that provides DC voltage at the end of electrodes. Improvement of this power supply type that has better purification and overall energy efficiency is obtained by the usage of intermittent supply [12].

5. Renewables and clean fuels

But, not only fossil fuel power plants affect on the environment. Renewable sources like small hydropower plants and wind farms could have significant influence on fish and bird habitats and migrations. The strategic environmental assessment can be considered as the most important, the most general and the most comprehensive instrument for directing the strategic planning process toward the principles and objectives of environmental protection, as well as for making optimum decisions on future sustainable spatial development, especially in energy sector [13].

Hydrogen is the most abundant element and cleanest fuel in the universe. Unlike hydrocarbon fuels that produce harmful emissions, hydrogen fuel produces pure water as the only by-product. Low-cost photoelectrochemical process efficiently uses sunlight to separate hydrogen from any source of water to produce clean and environmental friendly renewable hydrogen. Innovative solar hydrogen generator eliminates the need for conventional electrolyzers, which are expensive and energy intensive.

6. Conclusion

All of the above takes the attention of researchers to continuously work on solutions for better fuel usage and energy efficiency improvement, while producing more electricity with higher reliability and safety and lower impact to the environment. The aim of this book is to assist researches involved in power plant design and development, as well industrial engineers involved in plant's maintenance with recent techniques taken from different technologies and disciplines.

Author details

Aleksandar B. Nikolic* and Zarko S. Janda

*Address all correspondence to: anikolic@ieent.org

Electrical Engineering Institute Nikola Tesla, University of Belgrade, Belgrade, Serbia

References

[1] OECD. OECD Factbook 2015-2016: Economic, Environmental and Social Statistics. Paris: OECD Publishing; 2016. DOI: 10.1787/factbook-2015-en

[2] International Energy Agency. Key OECD Electricity Trends 2016. Available from: http://www.iea.org/media/statistics/Keyelectricitytrends2016_.pdf

[3] International Energy Agency. Key Trends on Energy Technology RD&D Budgets. 2016 Edition. Available from: http://www.iea.org/media/statistics/topics/IEA_RDD_Factsheet_2016.pdf

[4] Nikolic A, Pejovic B, Djuric B, Jankovic J, Drakic K. Maintenance improvement and cost reduction of large scale systems using remote monitoring and supervision. In: Proceedings of 2nd International Conference on Intelligent Control, Modelling and Systems Engineering (ICMS '14); 29-31 January 2014; Cambridge, MA, USA. pp. 229-235. WSEAS Press, 2014. ISSN: 2227-4588, ISBN: 978-960-474-365-0

[5] De Wachter B. Transformer Replacement Decisions. Application Note. ECI Publication No. Cu0185; November 2013

[6] Han Y, Song YH. Condition monitoring techniques for electrical equipment – A literature survey. IEEE Transactions on Power Delivery. 2003;**18**(1):4-13

[7] Kartalovic N, Babic B, Marinkovic S, Teslic D, Nikolic A. Monitoring and diagnostic center for generators. In: Proceedings of 2nd International Conference on Intelligent Control, Modeling and Systems Engineering (ICMS '14); 29-31 January 2014; Cambridge, MA, USA. pp. 151-155. WSEAS Press, 2014. ISSN: 2227-4588, ISBN: 978-960-474-365-0

[8] Hanna RA and Prabhu S. "Medium-voltage adjustable-speed drives-users' and manufacturers' experiences," in IEEE Transactions on Industry Applications, **33**(6):pp. 1407-1415, Nov/Dec 1997. doi: 10.1109/28.649949

[9] Janda Z, Nikolic A. MV variable speed drive for coal mill capacity improvement. In: Proceedings of 16th International Symposium on Power Electronics – Ee 2011; Paper No. T4-2.10, pp. 1-4. October 26th - 28th, 2011. Power Electronics Society, Novi Sad Serbia

[10] Dragosavac J, Janda Z, Milanovic JV, Mihailovic L, Radojicic B. Practical implementation of coordinated Q-V control in a multi-machine power plant. IEEE Transactions on Power Systems. 2014;**29**(6):2883-2891. DOI: 10.1109/TPWRS.2014.2318794

[11] United States Environmental Protection Agency. Climate Change Indicators in the United States: Global Greenhouse Gas Emissions. 2016. Available from: https://www.epa.gov/climate-indicators

[12] Parker K. Electrical Operation of Electrostatic Precipitators. London: The Institution of Electrical Engineers; 2003

[13] Nilssona M, Björklundb A, Finnveden G, Johanssonc J. Testing a SEA methodology for the energy sector: A waste incineration tax proposal. Environmental Impact Assessment Review. 2005;**25**:1-32. DOI: 10.1016/j.eiar.2004.04.003

Section 2

Power Plants Management, Maintenance and Reliability

Chapter 2

Key Technical Performance Indicators for Power Plants

Simona Vasilica Oprea and Adela Bâra

Additional information is available at the end of the chapter

http://dx.doi.org/10.5772/67858

Abstract

In this chapter, we will underline the importance of the key performance indicators (KPIs) computation for power plants' management. The main scope of the KPIs is to continuously monitor and improve the business and technological processes. Such indicators show the efficiency of a process or a system in relation with norms, targets or plans. They usually provide investors and stakeholders a better image regarding location, equipment technology, layout and design, solar and wind exposure in case of renewable energy sources and maintenance strategies. We will present the most important KPIs such as energy performance index, compensated performance ratio, power performance index, yield, and performance, and we will compare these KPIs in terms of relevance and propose a set of new KPIs relevant for maintenance activities. We will also present a case study of a business intelligence (BI) dashboard developed for renewable power plant operation in order to analyze the KPIs. The BI solution contains a data level for data management, an analytical model with KPI framework and forecasting methods based on artificial neural networks (ANN) for estimating the generated energy from renewable energy sources and an interactive dashboard for advanced analytics and decision support.

Keywords: Power plants, key performance indicators, renewables, business intelligence, forecasting models

1. Introduction

The main objective of key performance indicators (KPIs) evaluation and monitoring consists in detecting low performance in power plant operation, investigating issues and setting up maintenance plans in order to minimize the operational costs. Another objective is to point out the commissioning and inspection of power plants after major repairs so that the results recorded during a period of at least 6 months will be compared with the expected results from the climatic conditions, design and exposure point of view, etc. The objective entails identifying errors related to layout in case of renewables (especially photovoltaic power plants), incorrect installation, equipment failure, damage, premature aging, etc.

In order to provide a real time and complete analysis of KPIs, it is necessary to develop informatics systems that monitor and report the operational activity of the power plant and offers decision support for stakeholders. Various informatics solutions and applications are currently proposed and used, especially for renewable power plants' management: decision support systems (DSS) for wind power plants with (GIS) Geographic Information Systems capabilities [1], DSS for off-shore wind power plants [2] or GIS DSS for photovoltaic power plants [3]. Also, there are well-known software solutions for power plants' complete management provided by Siemens or Emerson that can be set up and customized depending on the equipment's configuration, location and size.

In this chapter, we will present the main key performance indicators for wind and photovoltaic power plants, identify new indicators for maintenance activities and propose an informatics solution that monitors and analyzes these KPIs through an interactive dashboard developed as a business intelligence portal accessed as a cloud computing service. The proposed solution is developed as part of the research project—intelligent system for predicting, analyzing and monitoring performance indicators and business processes in the field of renewable energies (SIPAMER), funded by National Authority for Scientific Research and Innovation, Romania, during 2014–2017.

2. Key performance indicators for power plant operation

The main objectives of assessing the technical performance of power plants based on renewable sources are

- Monitoring the operation of generating units or groups, identifying decline in their performance and also the need to perform maintenance/repairs on the affected groups. In this case, we recommend the use of energy performance index (EPI) and compensated performance ratio (CPR);
- Commissioning, recommissioning or evaluation after repairs, benchmarks for measuring and comparing further performance. We recommend using energy performance index (EPI) and power performance index (PPI);
- Calculating specific parameters such as yield, performance ratio (PR) to enable comparisons between power plants operation in different geographical areas and assisting decisions regarding investment in new groups or extending existing ones. In some cases, depending on the objectives, it is recommended to use several indicators (yield, PR, CPR, and/or EPI, depending on the level of effort and the level of uncertainty), so that the comparison to be more efficient.

http://dx.doi.org/10.5772/67858

Technical performance indicators allow the following comparisons:

- Operation of the power plant or a group compared with expectations at some different points in its runtime period;
- Operation of the power plant for a period of assessment compared to other power plant operation under similar climatic conditions;
- Standard power plant operation on short and long term in comparison with power plant operation under certain conditions (design, location, exposure, etc.);
- Power plant operation in consecutive time, the current performance being compared to past performance.

The main objective of the technical performance evaluation consists in detecting the decrease of power plant performance, investigating issues and completion of the maintenance operations, so that the involved costs are minimal.

In this section, we will present a series of key performance indicators for monitoring the operation of the wind power plants (WPP) and photovoltaic power plants (PPP). For a better analysis, we grouped KPIs in four categories: operational KPIs, indicators for photovoltaic power plants, indicators for wind power plants, and maintenance KPIs.

2.1. Performance indicator techniques based on operational data

1. The average power (P_{avg}) is the ratio between the produced energy (W) and power plant's runtime (t), depending on the yearly power plant operational time. According to [4], we may consider t as follows:

   - onshore WPP, t = 1900 hours/year;

   - offshore WPP, t = 3500 hours/year;

   - solar, t = 1100 hours/year.

$$P_{avg} = \frac{W}{t} \ [kW] \tag{1}$$

   The average power calculated at different time intervals is necessary to determine the installed power load factor. P_{avg} allows comparisons between monthly/quarterly or annual results of the same power plant, or it can be used to compare the generating units' performance within the same power plant.

2. Installed power load factor (K_u) is calculated as the ratio of average power (P_{avg}) and installed power (P_i):

$$K_u = \frac{P_{avg}}{P_i} \tag{2}$$

This coefficient can be calculated on monthly, quarterly or annually basis and indicates the availability of renewable resource and production capacity of the power plant. Also, it can indicate the degree of generating units or equipment's aging but must be correlated with meteorological factors that influence the production. For example, for wind power plants, the installed power load factor can range between 0.15 and 0.39.

3. Installed power load duration (T_i) is determined based on installed power load factor (K_u) multiply by power plant's runtime (t):

$$T_i = K_u \times t\ [h] \tag{3}$$

For photovoltaic power plants, the number of operating hours can be accordingly reduced, considering only those daytime hours when the PPP is operating. We may consider [4] for reference to operational time.

4. Maximum power load duration (T_{max}) is calculated as ratio between generated energy (W_a) and maximum power plant output (P_{max}):

$$T_{max} = \frac{W_a}{P_{max}}\ [h] \tag{4}$$

P_{max} can be calculated on monthly, quarterly or annual basis, and it can be used to compare results between different periods of time and identify the influence factors.

5. Power factor ($\cos\varphi$) can be determined based on active energy (W_a) and reactive energy (W_r):

$$\cos\varphi = \frac{1}{\sqrt{1 + \left(\frac{W_r}{W_a}\right)^2}} \tag{5}$$

Power factor is monitored for energy quality assurance.

6. Performance index (PI) is the ratio between the generated power/energy and forecasted power/energy:

$$\text{PI} = \frac{W}{W_f} \tag{6}$$

As described in [5], unlike performance ratio, index performance should be very close to 1 for the proper functioning of the PPP, and it should not vary from season to season due to temperature variations. There are several definitions of formal performance index:

- Energy performance index (EPI)—measures the energy (kWh) for a specific time period;

- Power performance index (PPI)—measures the effective power of the power plant (kW).

Energy or power forecast can be determined using different prediction models (regression model using historical data operation or system advisor model (SAM) which uses current climate data as input), thus the accuracy of performance index depends on the accuracy of the used forecast model. In Section 3, we will present a forecasting model based on artificial neural networks (ANN) for estimating the generated energy for photovoltaic and wind power plant.

2.2. Key performance indicators for photovoltaic power plants

Several technical performance indicators for PPP were defined by different organizations, for example, National Renewable Energy Laboratory (NREL) [6], the International Electrotechnical Commission (IEC) [7], associations and companies in the industry. Some of them are described in the following sections:

1. Performance ratio (*PR*) is defined according to IEC 61724 standard [7], as follows:

$$PR = \frac{Y_f}{Y_r} = \frac{\frac{kWh_{AC}}{kW_{DC\ STC}}}{\frac{kWh_{Sun}}{1kW}} \qquad (7)$$

Where:

- Y_f represents the ratio between annual active energy and rated power;

- Y_r is the ratio between insolation (kWh/m^2) and reference solar irradiance (1000 W/m^2). Irradiation is an instant size of solar power in a given area, and insolation measures energy gained for a certain area for a certain period of time.

Performance ratio can be evaluated on different time intervals (hourly, monthly, quarterly and annually). The main disadvantage of this indicator is that it is sensitive to temperature variations, and when plotted in a typical year, the index values are lower in warm periods and higher in cold periods.

It can be calculated on annual basis to make comparisons between photovoltaic power plants having similar climatic conditions but is not suitable for short periods of time or for comparing PPP efficiency under different climatic conditions.

2. Compensated performance ratio (CPR)

As reflected in the performance ratio formula, it is directly influenced by the energy produced by the photovoltaic power plant, which is directly influenced by solar irradiation and indirectly by the cell temperature. Therefore, it appears that PR decreases with increasing temperature.

According to [5, 8], offsetting factors such as cell temperature (K_{temp}) can be applied to the performance ratio to adjust the rated power under standard test conditions (STC).

$$PR_{TempComp} = \frac{\frac{kWh_{AC}}{kW_{DC\ STC} \times K_{Temp}}}{\frac{kWh_{Sun}}{1kW}} \qquad (8)$$

Where

- $K_{Temp} = T_{Cell} - T_{STC}$

- $T_{STC} = 25^{\circ}C$

This indicator is suitable for daytime values due to the fact that during night, the PPP production, irradiation and insolation are zero.

3. The yield is the ratio between the PPP's produced energy (kWh) during the operation time (*t*) and peak load power (kWp or kW peak) of the PPP or rated power on standard test conditions (STC), and it varies yearly depending on climate conditions.

The yield is determined annually based on the formula:

$$Yield = \frac{\sum_{i=1}^{t} kWh_{AC}}{kW_{DC\ STC}} \tag{9}$$

Due to the fact that the yield increases with the number of hours of operation and insolation etc., a high yield due to favorable climatic conditions can mask problems of premature aging of the equipment and vice versa.

When comparing the performance for two power plants or the yield for the same PPP in different periods of time, then the number of hours, insolation and cell temperature must be equivalent to achieve a fair comparison. Also, the power plant output (measured annually or at smaller intervals) can be compared with PPP's output from previous years. In this case, it must be taken into consideration the climate influence and correct the differences with a correction coefficient, to avoid masking problems of degradation of solar panels.

4. Normalized efficiency is another KPI for measuring the performance ratio [8]:

$$\eta_N = \frac{\frac{P}{P_n}}{\frac{E_{POA}}{E_{ref}}} \tag{10}$$

Where:

- P is the measured power;

- P_n is the rated power;

- E_{POA} is the plane-of-array irradiance;

- E_{ref} is the reference irradiation (1000 W/m^2).

Exposure to irradiation measures the total available solar exposure, and it is based on location exposure and direction of modules. It is calculated at the module level and average at central level. In order to maximize exposure to irradiation, modules are oriented towards the equator, the tilt modules depending on geographical latitude of the location. Optimal orientation in terms of space restrictions may not coincide with the orientation that maximizes exposure (due to the fact that a lower slope leads to more modules in a project).

One drawback of the performance index is that the normalized efficiency is sensitive to temperature variations, as any change in temperature leads to changes in efficiency, power and consequently in the produced energy.

Changing efficiency or power for a photovoltaic module can be quantified using the temperature coefficient of power γ, which allows the module power (or efficiency) to be modelled to a certain temperature. For silicon crystals, γ is between $-0.3\%/^{\circ}C$ (for newer technologies) and $-0.5\%/^{\circ}C$ (for older technologies).

Power for a certain temperature for a photovoltaic cell is determined by:

$$P(T) = P_{STC}\left(1 + \gamma(T_{cell} - T_{STC})\right) = P_{STC}(1 + \gamma\Delta T_{STC}) \tag{11}$$

Where

- T_{STC} is 25°C;

- T_{cell} is the temperature of the photovoltaic cell.

The temperature—corrected power (P^*) can be determined as in [9]:

$$P^* = \frac{P(T)}{1 + \gamma\Delta T_{STC}} \tag{12}$$

Thus, the temperature-corrected normalized efficiency can be expressed as:

$$\eta_N{}^* = \frac{\frac{P^*}{P_n}}{\frac{E_{POA}}{E_{ref}}} \tag{13}$$

This indicator shows the performance of the photovoltaic module as if it operates at standard temperature (T_{STC}). In this way, the technical performance of the PPP can be attributed to other factors, such as the irradiance spectrum or inverter efficiency at lower irradiances [9].

2.3. Key performance indicators for wind power plants

1. Specific energy production (SPE) measured in kWh/m^2 for a wind turbine is defined in [10] as the ratio between total energy production during nominal period (W) and swept rotor area (S_{SR}):

$$SPE = \frac{W}{S_{SR}} \tag{14}$$

The nominal period is the period covered by the report, usually considered as 1 year. SPE is also called as energy yield or energy productivity [11], and it depends on the turbines' rated power.

2. Capacity (load) factor (CF %) defined in [11] is the ratio between total energy production during the nominal period (W) and the potential energy production during the reported period (W_p):

$$CF = \frac{W}{W_p} \tag{15}$$

Usually, the capacity factor varies depending on the turbines specifications and climate conditions between 18 and 40% for onshore turbines and 30–40% for offshore turbines.

3. Equivalent full load hours (E_h) can be defined as the annual energy production (W) divided by the rated power (S_n), and it represents the number of hours as if turbines generate at rated power:

$$E_h = \frac{W}{S_n} \tag{16}$$

4. Availability factor (%) represents an important indicator especially for WPP due to the wind influence that affects the turbines' generation and can be calculated as ratio between total hours of operation during the reported period (T_{op}) and total hours of reported period (T_p):

$$AF = \frac{T_{op}}{T_p} \tag{17}$$

2.4. KPIs for maintenance operations

Several maintenance strategies have been developed as described in [12, 13] with the main objective to preserve the efficiency of power plants' components. Each of these methodologies has its own characteristics, but mainly they focus on internal characteristics of the power plants' components. The industry has adopted for a long period of time maintenance that focuses on corrective actions. But, in recent years, the maintenance plans focus on predictive maintenance where monitoring or inspection activities are performed to determine the best time to start the maintenance in order to minimize the efforts compared to corrective maintenance.

Preventive maintenance activity has a direct impact on the reliability of the equipment or components by improving their technical condition and prolonging their life. All maintenance procedures involve both costs and benefits. Maintenance operations are profitable when the costs are lower than associated potential cost of a failure, which these operations are trying to prevent. Most of the maintenance plans on short and medium term do not take into account the operation conditions in which the components operated throughout their runtime but rather are scheduled based on the occurrence of defects and previous repairs. But, in recent years, several applications for continuous monitoring of current operation led to the development of a variety of diagnostic techniques. According to [14], these techniques verify certain parameters and then analyze whether certain components are defective at the moment and can make an estimate of their evolution.

The main purpose of the maintenance plan is to minimize production costs per unit of energy generated. In general, this is achieved by minimizing operational and maintenance costs, improving turbine/photovoltaic panels' performance and efficiency and lowering insurance policy and equipment's protection. Thus, we proposed two KPIs for determine the loss due to preventive (planned) maintenance or to corrective (unplanned) maintenance.

1. Preventive loss indicator (PLI_{plan}) is the ratio between estimated energy loss caused by planned interruptions and the maximum energy that can be produced during the reported period (usually 1 year).

$$PLI_{plan} = \frac{W_{lossplan}}{W_{max\_prod}} \times 100\% \tag{18}$$

Where:

- $W_{lossplan}$ represents the energy loss caused by the planned interruptions;

- W_{max_prod} is the maximum energy that can be produced during the reported period.

2. Corrective loss indicator (PLI_{unplan}) is the ratio between estimated energy loss caused by unplanned interruptions and the maximum energy that can be produced during the reported period.

$$PLI_{unplan} = \frac{W_{lossunplan}}{W_{max\_prod}} \times 100\% \tag{19}$$

Where:
- $W_{lossunplan}$ represents the energy loss caused by the unplanned interruptions;

- W_{max_prod} is the maximum energy that can be produced during the reported period.

Depending on these indicators, the maintenance policy can be schedule in order to minimize the production losses.

3. Informatics solutions for monitoring and analyzing the power plants' KPIs

In order to analyze and monitor the key performance indicators, the executives of the power plants require an advanced decision support system (DSS). Our proposal consists in developing an informatics solution based on three levels architecture that involves models for data management, analytical models and interfaces (**Figure 1**):

The architecture components are as follows:

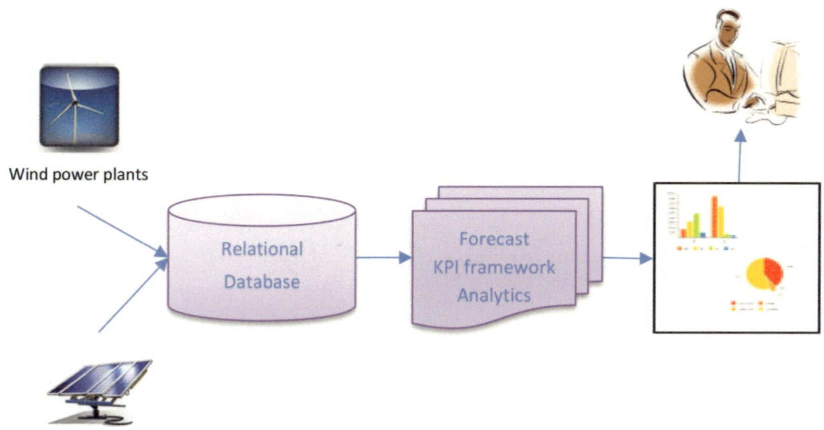

Figure 1. SIPAMER's architecture.

3.1. Level 1—data management

All data sources gathered from wind/photovoltaic power plants are extracted, transformed and loaded into a central relational database running Oracle Database 12c Edition in order to enable user access through cloud computing. The sources are heterogeneous: measuring devices for climate conditions (wind speed, direction, temperature, atmospheric pressure, and humidity), sensors for photovoltaic cells and wind turbines, SCADA API for measuring real-time parameters regarding power plant output. These sources are mapped into a relational data stage; then, the extract, transform and load (ETL) process is applied, and data are finally loaded into a relational data mart that organizes objects as dimensions and facts. This approach makes it easier the development of the analytical model with KPIs framework and enables an advanced roll-up/drill-down interfaces.

Based on the executives' requirements regarding the KPIs, we designed the main structural entities (objects) that will enable multidimensional data exploration. They will be organized as dimensions (subject entities) with descriptive attributes structured on hierarchies with multiple levels to enable typical OLAP operations: roll-up/drill-down, slicing and dicing. The data mart contains the following dimensions: DIM_STAKEHOLDER, DIM_POWERPLANT, DIM_REGION, DIM_TURBINE, DIM_PV and DIM_TIME.

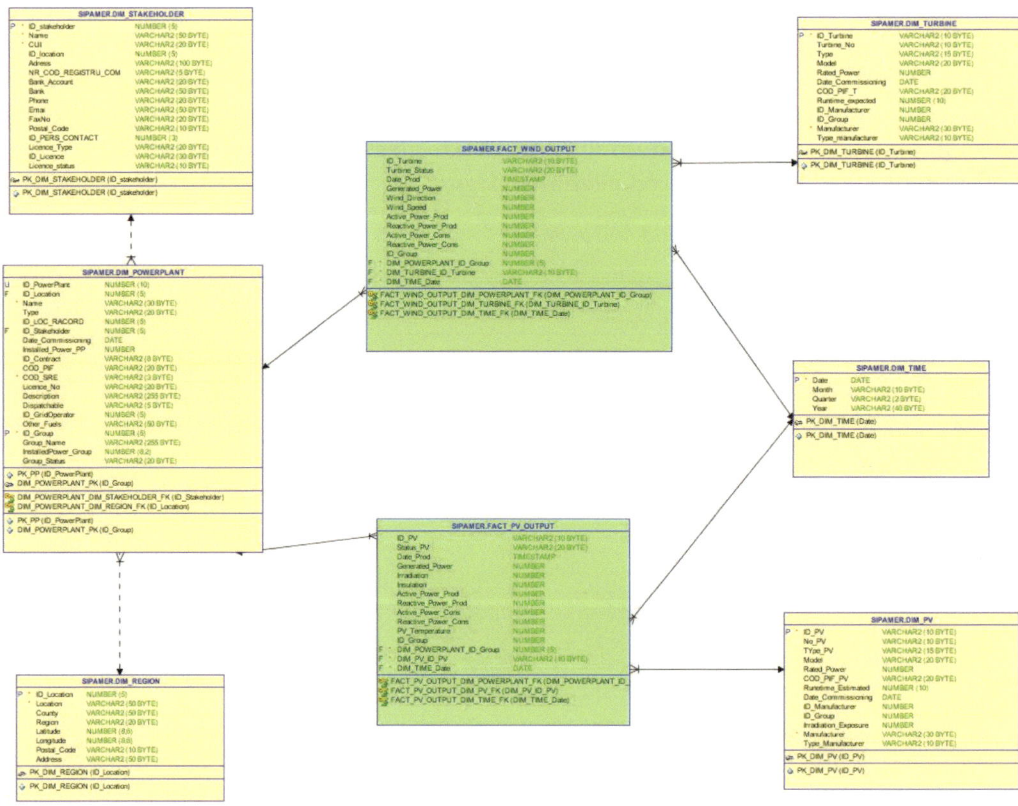

Figure 2. Snowflake schema for the KPIs data mart.

http://dx.doi.org/10.5772/67858

Facts tables are objects that contain attributes like measures (metrics) and foreign keys to the dimension tables. Facts are usually numerical data that can be aggregated and analyzed by dimensions' levels. The model contains the following facts: FACT_PV_OUTPUT and FACT_WIND_OUTPUT. The objects are organized in a snowflake schema as shown in **Figure 2**.

The data mart allows us to design the KPIs framework in a subject-oriented and multidimensional view.

3.2. Level 2—models

This level contains models for forecasting the power plant output on short term (hourly, up to 3 days) and the KPIs analytical framework.

Forecasting models are build distinct for each type of renewable power plant, WPP and PPP due to the different influence factors that affect the power plant' operation and generation. The aim of the model is to improve predictions made and transmitted currently by the producer on short-time intervals. The deviations between forecasting and recorded production are currently about 30–35% for wind power plants and 15–20% for photovoltaic power plants [15, 16]. Minimizing these deviations will lead to lower costs for stakeholders due to the fact that imbalances are paid. The model consists in a set of experimental methods based on data mining algorithms, developed, validated and tested on WPP and PPP data sets. We developed three algorithms based on artificial neural networks (ANN): Levenberg-Marquardt algorithm (LM), Bayesian regularization algorithm (BR), and scaled conjugate gradient algorithm (SCG).

3.2.1. Forecasting the photovoltaic power plants' output

We identified the input parameters (irradiance, temperature, wind speed & direction, tilt, exposure) and the output (power), and for the training and validation, we used a data set that consist of 50,631 samples from every 10 minutes direct measurements in a PPP located in Romania, Giurgiu County, during January 1, 2014—December 31, 2014. Within this photovoltaic power plant are installed two types of ABB—PSV800 invertors, with 600 kW and 760 kW, 30,888 solar panels and the solar module has a rated power of 245 W with a 20-kV connection. The configuration is widely used in other PPP; therefore, the developed ANN can be easily implemented in other power plants with similar configuration.

Since solar energy presents seasonal variations related to the various climate conditions of the year, we designed the neural networks adaptable to irregular seasonal variations by changing the settings on the number of neurons in hidden layers and developed two types of ANN.

First, we designed one neural network for each of the three algorithms (LM, BR and SCG) based on the whole year data. The results were good, with an average mean squared error (MSE) of 0.19, and average for correlation coefficient, $R = 0.95$, with 0.9573 for LM.

Then, we consider the second option, to take into account the seasonal variations for solar energy, and we designed neural networks based on LM, BR and SCG for each month. So, we obtained 36 neural networks with a much better results than the previous case (yearly ANNs). Comparing results from the monthly data, we found that the prediction accuracy is excellent in

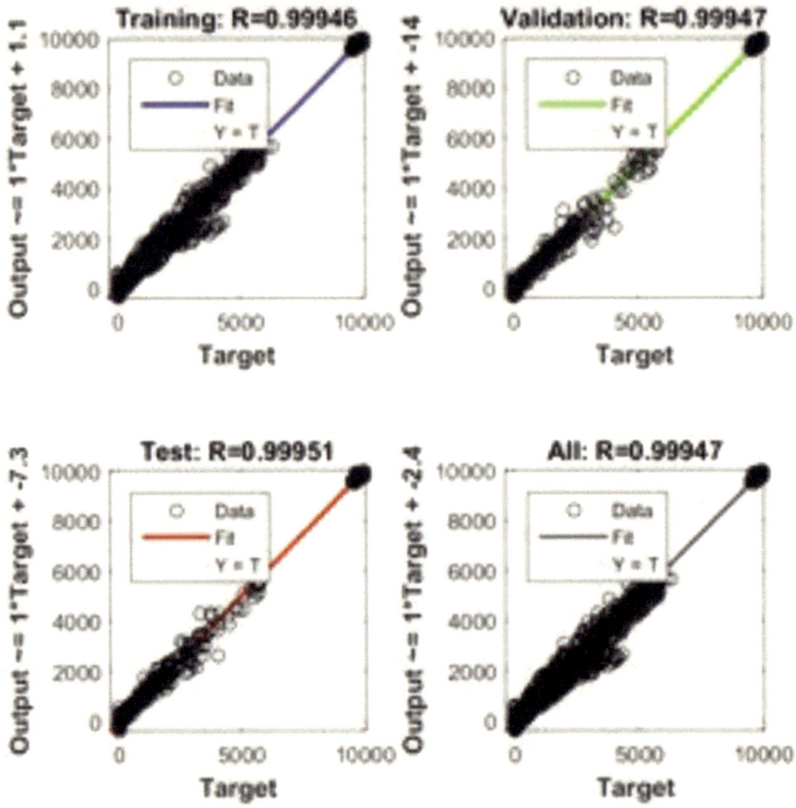

Figure 3. Regression between target values and the output values of the neural network *SfebruaryLM*.

all months, and monthly performance indicators have comparable values. The MSE is between 0.03 and 0.1, and coefficient R is between 0.997 and 0.999. For example, **Figure 3** shows the correlation coefficient for the neural network *SFebruaryLM* developed on Levenberg-Marquardt algorithm.

By comparing the forecasting results through the development of neural networks based on the three algorithms, we found that in 69% of cases, neural networks developed with Bayesian regularization produced a better generalization than networks developed with Levenberg-Marquardt and SCG algorithms. But, in 31% of cases, the forecasting results with the highest level of accuracy have been obtained in the case of Levenberg-Marquardt algorithm.

If, in order to improve the accuracy of the forecasting model, new elements are added as input data, the LM algorithm will offer the advantage of a higher training rate compared with the BR algorithm but would have the disadvantage of an increased memory consumption. When new inputs are added and we want to obtain a high speed and performance, then the best solution is to develop the ANN based on SCG algorithm as it is faster than the other two algorithms (LM and BR) requiring low memory consumption, with the drawback that it provides a lower level of prediction accuracy.

http://dx.doi.org/10.5772/67858

3.2.2. Forecasting the wind power plants' output

We identified the input parameters (temperature, wind speed & direction at 50 m, 55 m, 75 m, 90 m, humidity, atmospheric pressure, turbine height, soil orography, slipstream effect) and the output (power). For ANN training and validation, we used a data set of 17,491 samples from hourly measurements in a WPP located in Romania, Tulcea, for 2 years (January 1, 2013–December 31, 2014). In this WPP, there are two types of wind turbines: V90 2MW/3MW IEC IA/IIA, with a height of 90 meters. These types of wind turbines are commonly used, so we can consider the data set suitable for training a generalized neural network, as described in [17].

Since wind energy presents seasonal variations over 1 year period, we design two sets of ANN based of three algorithms: Levenberg-Marquardt algorithm (LM), Bayesian regularization algorithm (BR) and scaled conjugate gradient algorithm (SCG).

First, we designed the neural network based on data set covering 2 years records for each algorithm (LM, BR and SCG). For the second solution, we take into account seasonal variations that affect wind energy and designed neural networks for each season, dividing the data into 4 sets corresponding to 4 seasons specific to Romania region. The results between the ANN trained for the whole year and the ANN trained for corresponding season are compared in **Table 1**.

The best approach is to develop and train the neural networks adjusted with seasonal data, due to the fact that the prediction accuracy is excellent in all seasons, and performance indicators have comparable values. Comparing the results for each algorithm (LM, BR, SCG), in most cases, neural networks based on Bayesian regularization produced a better generalization than Levenberg-Marquardt or SCG algorithms, but LM performed faster and with minimum memory consumption.

KPIs analytical framework provides methods for calculating the key performance indicators used by executives to monitor the power plants in terms of technological and business processes. For technological processes, we build the KPIs presented in Section 2 based on formulas (1) to (19). For business processes, we included commonly used KPIs as income, cost, profit/loss, etc. The KPIs are developed directly into the facts tables, as derived measures and accessible into the interface level.

Period	MSE			R			Errors interval		
	LM	BR	SCG	LM	BR	SCG	LM	BR	SCG
Year	0.06090	0.05789	0.06640	0.92922	0.93306	0.92739	−0.9651; 0.8758	−1.001; 0.8677	−0.7543; 0.7062
Spring	0.04817	0.03608	0.05079	0.96238	0.95877	0.94980	−0.3296; 0.2347	−0.3811; 0.1868	−0.2089; 0.3043
Summer	0.06692	0.05970	0.05985	0.95232	0.95695	0.93996	−0.2564; 0.1772	−0.2558; 0.1779	−0.3218; 0.28
Autumn	0.07313	0.06837	0.11224	0.92163	0.93524	0.93134	−0.6642; 0.5724	−0.2466; 0.2653	−0.4059; 0.261
Winter	0.05927	0.05591	0.06317	0.95109	0.95147	0.94960	−0.3932; 0.5173	−0.3445; 0.4948	−0.3835; 0.5423

Table 1. Comparison between ANN developed for one year and ANN with seasonal adjustments.

3.3. Level 3—interface

The forecasting and analytical models are integrated into an online dashboard developed in Java with application development framework (ADF). The dashboard is built as a business intelligence (BI) portal with a very friendly interface and interactive charts, reports, pivot tables, maps and narrative elements that allows executives and stakeholders to easily analyze the KPIs. The dashboard contains three sections:

- Production management—it contains reports for power plant current operations and maintenance plans, it displays the generation groups' configuration and location, real time data gathered from measuring devices, SCADA and generation groups or the entire power plant;
- Forecasting—it contains access to the forecasting models and offers reports and charts to display the estimations versus actual values for different periods of time, selected by the user. For example, **Figure 4** shows a chart that displays for one day interval, on hourly

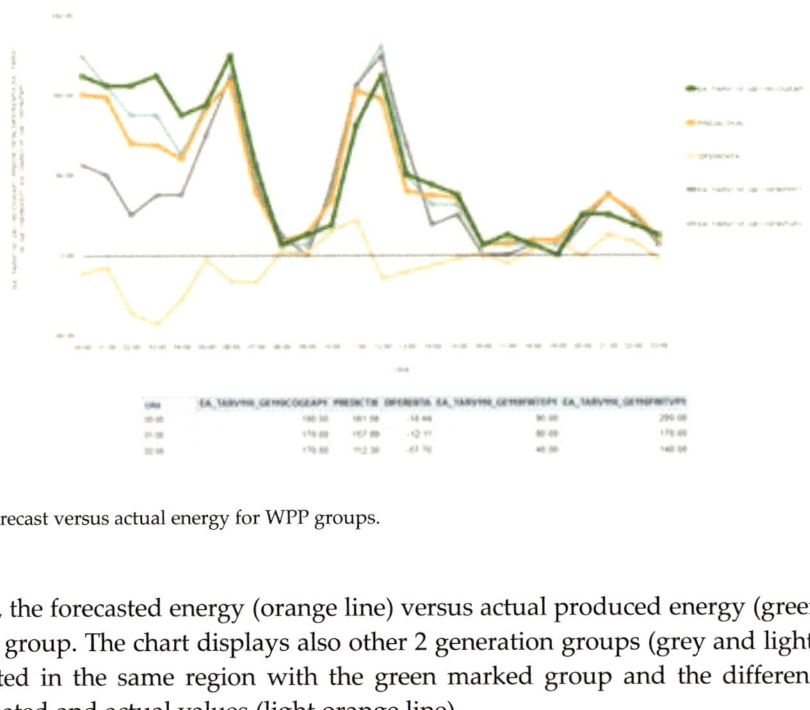

Figure 4. Forecast versus actual energy for WPP groups.

basis, the forecasted energy (orange line) versus actual produced energy (green line) for a WPP group. The chart displays also other 2 generation groups (grey and light blue lines) situated in the same region with the green marked group and the difference between estimated and actual values (light orange line).

- KPIs Analytics—contains analytical Business Intelligence elements (interactive charts, gauges, reports, maps, pivot tables) that enable KPI advanced analysis through dimensions' hierarchies that allows executives to compare indicators over different periods of time, regions and locations, aggregate/detailed KPIs over power plants' groups or module/turbines. For example, **Figure 5** shows the average power, installed power load factor, installed power load duration and maximum power load duration for a wind power plant with two groups of 5 and 10 MW.

http://dx.doi.org/10.5772/67858

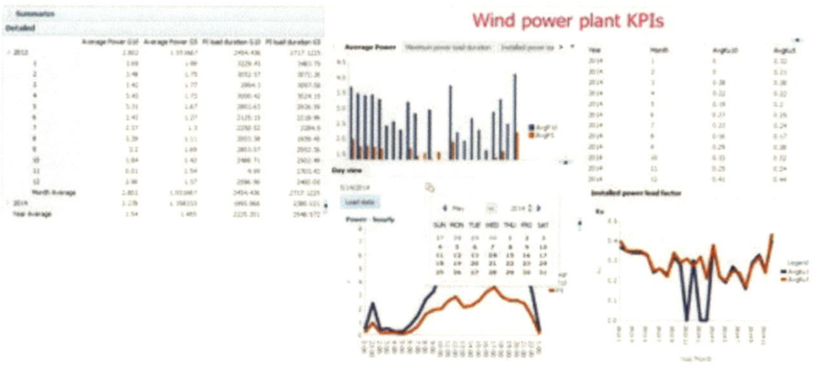

Figure 5. KPIs dashboard.

The dashboard is developed in a cloud computing architecture, and it is accessible as a service, customized and configured depending on stakeholders' interest.

4. Conclusions

In this study, we proposed a framework for calculating the most relevant key performance indicators for wind power plants and photovoltaic power plants that offer a realistic perspective on technical aspects of the operational and maintenance activities. Also, it is proposed an informatics solution for KPIs analysis that can support decision process and integrates models for data management, analytical models and interactive interfaces.

Through the business intelligence dashboard that integrates the key performance indicators, the stakeholders can monitor the current operation of power plant and identify the decline in performance and the need to set up the maintenance strategy. Also, the KPI framework is useful for commissioning, recommissioning or evaluation after major repairs, establish benchmarks for measuring and comparing further performance.

The proposed solution integrates two major components: the forecasting model that provides estimations regarding the wind power plants' or photovoltaic power plants' output with a good accuracy for short-term interval (intraday and up to 3 days); the KPIs analytical model that allows a very interactive analysis of power plant management regarding past operation, detecting possible issues, offering smart analyses of KPIs, setting thresholds for metrics and present them in a user friendly and interactive dashboard.

Acknowledgements

This paper presents some results of the research project: Intelligent system for predicting, analyzing and monitoring performance indicators and business processes in the field of renewable energies (SIPAMER), PNII—PCCA 2013, code 0996, no. 49/2014 funded by National Authority for Scientific Research and Innovation, Romania.

Author details

Simona Vasilica Oprea and Adela Bâra*

*Address all correspondence to: bara.adela@ie.ase.ro

The Bucharest University of Economic Studies, Romania

References

[1] R. Maria, L. Bottaia, C. Busilloa, F. Calastrinib, B. Gozzinib, G. Gualtier. A GIS-based interactive web decision support system for planning wind farms in Tuscany (Italy). Renewable Energy. February 2011. Vol. 36 (2).

[2] T. Pahlke. Software & Decision Support Systems for Offshore Wind Energy Exploitation in the North Sea Region. POWER Project. [Internet]. 2007. Available from: http://pcoe.nl/@api/deki/files/1900/=12wp1_executivesummary_sdss-studie_2007-06-05.pdf [Accessed: 2016].

[3] E.B. Mondino, E. Fabrizio, R. Chiabrano. Site selection of large ground-mounted photovoltaic plants: a GIS decision support system and an application to Italy. International Journal of Green Energy. 2015. Vol. 12 (5).

[4] ENTSO-E Scenario Outlook & Adequacy Forecast 2013—2030, pp. 122 [Internet]. 2013. Available from: https://www.entsoe.eu/fileadmin/user_upload/_library/publications/entsoe/So_AF_2013-2030/130403_SOAF_2013-2030_final.pdf [Accessed: 2017].

[5] J. Mokri, J. Cunningham. PV System Performance Assesment. SunSpec Alliance, USA, 2014.

[6] National Renewable Energy Laboratory. http://www.nrel.gov/ [Internet]. [Accessed: December 2016].

[7] International Electrotechnical Commission. International Standard IEC 61724 [Internet]. 1998. Available from: https://webstore.iec.ch/preview/info_iec61724%7Bed1.0%7Den.pdf [Accessed: 2016].

[8] M. Taylor, D. Williams. PV Performance Guarantees (Part 1) Managing Risks & Expectations. SolarPro., USA, 2011.

[9] Sandia National Laboratories. https://pvpmc.sandia.gov/modeling-steps/5-ac-system-output/pv-performance-metrics/normalized-efficiency/[Internet]. 2014. [Accessed: December 2016].

[10] Public Utilities Commission of Sri Lanka. Performance Measurements of Generation and Transmission Systems [Internet]. 2014. Available from: http://www.pucsl.gov.lk/english/wp-content/uploads/2014/03/Performance-Indicators-Report.pdf [Accessed: 2016].

[11] World Energy Council. Technical Performance Indicators for Wind Energy [Internet]. 2012. Available from: https://www.worldenergy.org/wp-content/uploads/2012/10/PUB_Performance_of_Generating_Plant_Working_Group3Appendix_2007_WEC.pdf [Accessed: 2016].

[12] T. Knill, A. Oakey. Operation and Maintenance of Wind Farms—Introduction and Overview [Internet]. Available from: http://www.wwindea.org/technology/ch03/en/3_1_1.html [Accessed: 2016].

[13] K. Fischer. Maintenance Management of Wind Power Systems by means of Reliability-Centred Maintenance and Condition Monitoring Systems. Göteborg, Suedia: Chalmers University of Technology; 2012.

[14] M.A. Sanz-Bobi. Use, Operation and Maintenance of Renewable Energy Systems. Green Energy and Technology. Switzerland: Springer International Publishing; 2014. DOI: 10.1007/978-3-319-03224-5_2.

[15] A. Bâra, G. Căruţaşu, C. Botezatu, A. Pîrjan. Comparative Analysis between Wind and Solar Forecasting Methods Using Artificial Neural Networks. Proceedings of the 16th IEEE International Symposium on Computational Intelligence and Informatics (CINTI 2015). 19–21 November 2015. Budapest, Hungary. ISBN 978-1-4673-8520-6.

[16] A. Bara, I. Lungu, S.V. Oprea, G. Carutasu, C. Botezatu, C.P. Botezatu. Design workflow for cloud service information system for integration and knowledge management based in renewable energy. Journal of Information Systems & Operations Management. 2014. Vol. 8 (2). ISSN 1843–4711.

[17] I. Lungu, G. Căruţaşu, A. Pîrjan, S.V. Oprea, A. Bâra. A two-step forecasting solution and upscaling technique for small size wind farms located on hilly terrain in Romania. Studies in Informatics and Control Journal. 2016. Vol. 25 (1). ISSN 1220–1766.

Chapter 3

Predictive Maintenance Based on Control Charts Applied at Thermoelectric Power Plant

Emilija Kisić, Željko Đurović and Vera Petrović

Additional information is available at the end of the chapter

http://dx.doi.org/10.5772/intechopen.68685

Abstract

In this chapter, innovative predictive maintenance technique is described with the aim of highlighting the benefits of predictive maintenance compared to time-based maintenance. The proposed technique is applied to a specific problem that occurs when time-based maintenance is applied on grinding tables of the coal mill, in coal-grinding subsystem at the thermoelectric power plant 'TEKO', Kostolac, Serbia. Time-based maintenance provides replacement of grinding tables after certain number of working hours, but depending on the quality of the coal and grinding table itself, this replacement sometimes needs to be made before or after planned replacement. The consequences of such maintenance are great material losses incurred because of frequent shutdowns of the entire coal-grinding subsystem, as well as the possibility that the failure occurs before replacement. Innovative predictive maintenance technique described in the chapter is used for solution of this problem.

Keywords: predictive maintenance, T^2 control chart, hidden Markov model, thermoelectric power plant, statistical process control

1. Introduction

In today's industry, application of the best maintenance strategies is a very important task in ensuring stability and reliability of technical systems. Numerous papers and books about different maintenance strategies can be found in literature, and almost everywhere the merits of predictive maintenance in regard to time-based maintenance are emphasized [1]. Predictive maintenance extends the period of time during which the system functions well, decreases unnecessary shutdowns, reduces material losses and prevents catastrophic failures. Although this field of research is very much advanced with the development of highly sophisticated technologies, there is still a lot of room for improvement of the existing techniques and the development of new ones.

In this research, an innovative technique of predictive maintenance is proposed and applied to a specific problem that occurs at the thermoelectric power plant 'TEKO', Kostolac, Serbia. Namely, one of the key thermoelectric power plant components is the coal-grinding subsystem. When time-based maintenance is applied on grinding tables of the coal mill, grinding tables are replaced after certain number of working hours. Depending on the quality of the coal and grinding table itself, this replacement sometimes needs to be made before or after planned replacement. The only way to determine the condition of the grinding table is visual inspection, which implies the shutting down of the whole subsystem. Consequences of grinding table replacement after fixed time intervals are great material losses incurred because of frequent shutdowns of the entire coal-grinding subsystem. Also, there is a possibility that the failure will occur before replacement.

There is an 'urban legend' that experienced operators in industrial plants, such as thermoelectric power plants, can 'hear' the sounds in sound content from operational drives. Based on these sounds, they can recognize the detritions of specific elements that can wear out, such as mill-grinding tables. Also, in literature one can find that 99% of mechanical failures are foregone by some very noticeable indicators [2]. Because of these facts, the idea came up for the recording of acoustic signals while coal-grinding subsystem is operational. In this way, it is easy to obtain condition-monitoring data which can be applied for predictive maintenance, and there is no need for shutting down the whole subsystem for obtaining the information about grinding table condition.

The proposed method is a trade-off between solutions already offered in the literature, and originality of the proposed algorithm is based on the selection of failure prognostic technique. The main goal of the proposed algorithm is the increase of energy efficiency at the thermoelectric power plant.

This chapter is organized as follows: In the next section, we describe the concept of predictive maintenance in detail. In Section 3, a description of the coal-grinding subsystem in thermoelectric power plant will be given. In Section 4, we present a new predictive maintenance technique. Section 5 contains the results. The last section is the conclusion, with the discussion about gain results.

2. Predictive maintenance

Nowadays, industrial processes are very complex and cannot be imagined without modern technologies, so highly sophisticated and very expensive maintenance strategies are needed. Consequences of inefficient maintenance are large material losses, and because of that it is necessary to constantly develop and improve the existing maintenance programmes.

Maintenance strategies were evolving during time. The first maintenance strategy was the *unplanned maintenance* or *run-to-failure* maintenance which implies waiting for failure to occur. It is obvious that with this maintenance strategy catastrophic failures are unavoidable, so very rare this kind of maintenance is sustainable and profitable. Later, *preventive maintenance* was introduced. Preventive maintenance can be conducted as *planned maintenance* or *time-based maintenance*, which is implemented at fixed time intervals, or it can be conducted as *predictive maintenance* or *condition-based maintenance* where maintenance activities are realized based on the condition of the system. Although with time-based maintenance equipment failures sometimes can be reduced, it does not eliminate catastrophic failures and causes unnecessary maintenance. In literature, it can be found that in the USA, because of ineffective maintenance, more than 60 billion of dollars are spent every year [3]. Similar situation is in other countries. Namely, the biggest shortcoming of time-based maintenance is too often replacement of system's parts, as well as premature stopping of the system while it is operational, which leads to great material losses. In most situations, predictive maintenance is the best choice, especially when maintenance is very expensive and occurring of failure is unacceptable. The main goal of predictive maintenance is extension of time in which system functions well and at the same time reduction of unnecessary stoppages and failures. Also, the aim of predictive maintenance is to prevent the occurring of catastrophic failures which can produce not only material costs but also loss of lives and environment pollution. List of this kind of accidents is not small and can be found in Ref. [4]. Because of these catastrophic failures which occasionally occur in modern industries, more attention is paid to the improvement of the existing predictive maintenance strategies, as well as to introducing the new ones. If it is regularly established and effectively implemented, predictive maintenance can significantly reduce maintenance expenses through cutting down of unnecessary time-based maintenance operations [5].

Diagnostics and prognostics are two very important aspects in predictive maintenance programme. Diagnostics deals with fault detection, isolation and identification after occurring of the fault. Fault detection indicates when something goes wrong in a monitored system, that is, it detects that failure has occurred. Fault isolation has a task to locate faulty component, whereas fault identification has a task to determine the nature of the fault when the fault is detected. Diagnostics has been developed for years, and today it presents very important area in engineering and automatic control [6, 7].

Prognostics deals with fault prediction, before the fault will occur. In other words, diagnostics is the posterior analysis of events, while prognostic is a priori analysis of events. Prognostics is more efficient in regard to diagnostics for achieving zero-downtime performances. On the other hand, diagnostics is necessary when failure prediction within prognostic fails and fault occurs. References about prognostics can be found in Refs. [8, 9]. Predictive maintenance can be used for diagnostics and prognostics, or both. Some newer references about predictive maintenance can be found in Refs. [10–12]. No matter what is the goal of predictive maintenance, three key steps must be followed for its implementation: (1) data acquisition, (2) data processing and (3) maintenance decision-making.

Data acquisition is the process of data collection from specific physical resources in order to implement predictive maintenance properly. This process is the key step in applying predictive maintenance, both for diagnostics and for prognostics. Collected data can be classified into two major categories: *event data* and *condition-monitoring data*. Event data include information about what happened (faults, repairs, what were the causes, etc.). Condition-monitoring data are the measurements about physical resource 'health condition'.

The first step in data processing is data cleaning. This step is very important, because data (especially event data), which are entered manually always, have some mistakes. Without data cleaning, it is possible that diagnostics and prognostics will be inaccurate. The next step in data processing is data analysis. Different models, algorithms and tools for data analysis depend mostly from data type [5]. Condition-monitoring data can be classified into three categories: (1) value type, (2) waveform type and (3) multidimensional type.

The last step in predictive maintenance programme is decision-making. Techniques for decision-making can be divided into two categories: *diagnostics* and *prognostics*. It is obvious that prognostics is superior in regard to diagnostics, because it can prevent failure to occur, and if it is possible it provides spare parts and planned human resources for problems that will occur. In this way, it is possible to reduce material losses and avoid catastrophic failures. However, prognostics cannot replace diagnostics completely, because in practice there will be always some unpredictable faults.

Here, we focus on prognostics. There are two types of prediction when we talk about *failure prognostic*. The first type is the prediction of how much time is left before failure will occur (one or more failures) depending on the current state of the machine and past operation profile. Time that is left before the fault is noticed is called *remaining useful life* (*RUL*). In some situations, especially when failure is catastrophic (e.g. nuclear plant), it is much a preferable second type of failure prognostic, that is, prediction of probability that the machine will work until some future time (e.g. until next interval when inspection is needed) depending on the current state of the machine and past operation profile. Actually, in any situation, it is good to know the probability that a machine will work without failure until the next inspection or condition monitoring. Most papers deal with the first type of failure prognostic, that is, with *RUL* estimation [13, 14]. Only few papers can be found that deal with the second type of prognostic [15]. According to Ref. [8], failure prediction can be divided into three different categories:

1. Traditional reliability approaches—prediction based on event data (experience) [16]
2. Prognostics approaches—prediction based on condition-monitoring data [17, 18]
3. Integrated approaches—prediction based on event data and condition-monitoring data [19].

Every one of these approaches has some advantages and limitations. Combinations of these approaches are different according to their applicability, price, precision and complexity [20].

3. Description of the coal-grinding subsystem in thermoelectric power plant

Thermoelectric power plants are the largest producers of electricity in Serbia, contributing with more than 65% of the total electricity supply. In order to ensure their stability and operational efficiency, it is necessary to monitor their major subsystems and individual components. In this way, it is possible to detect any change in behaviour, or failure on time, which leads to the increase of energy efficiency and the reduction of the financial losses of the electric power industry.

http://dx.doi.org/10.5772/intechopen.68685

One of the key thermoelectric power plant components is the coal-grinding subsystem. Its physical layout is shown in **Figure 1**. Raw coal enters the subsystem through a feeder and goes down a chute to the grinding table that rotates at a constant speed. The coal is then moved outward by centrifugal force and goes under three stationary rollers where it is ground. The outgoing coal moves forward to the mill throat where it is mixed with hot primary air. The heavier coal particles immediately move back to the grinding table for additional grinding, while lighter particles are carried by the air flow to the separator. The separator contains a large amount of particles suspended in the powerful air flow. Additionally, some of the particles drawn into the primary air-and-coal mix lose their velocity and fall onto the grinding table (as shown) for further grinding, while the particles that are fast enough enter the classifier zone. These particles are swirled by deflector plates. Lighter particles are removed as classified fuel in the form of fine powder that goes to burners, while heavier particles bounce off the classifier cone

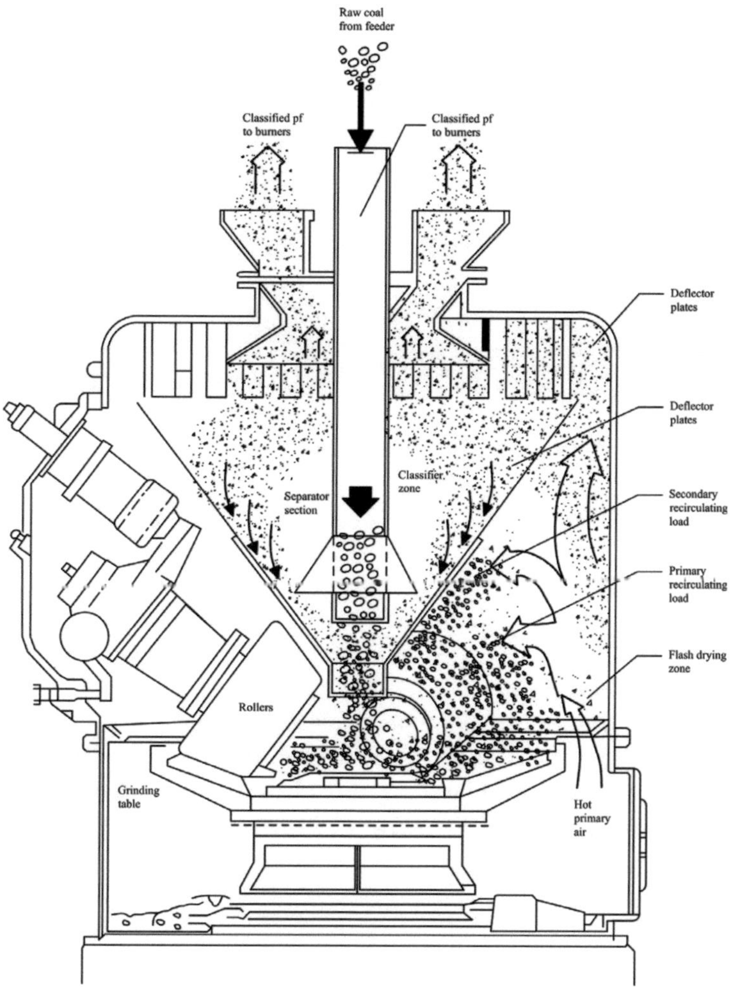

Figure 1. Configuration of the coal-grinding subsystem.

and fall back onto the grinding table for additional grinding. Both the separator and classifier contain a significant amount of coal. These coal masses, along with the coal on the grinding table and the three recirculating loads (primary, secondary and tertiary), play a key role in the dynamic performance of the mill [21, 22].

In this research, one such system at the thermoelectric power plant 'TEKO' (Serbia) is analysed. As it is previously described, the coal inside the mill is ground by impact and friction against the grinding table that rotates around the mill centre line (CL). The only way to determine the current condition of the grinding table is to shut down the entire subsystem and open it for visual inspection. This time-based maintenance method guarantees that grinding tables will be replaced before they become dysfunctional, but at a cost of frequent shutdowns. If inspection shows that grinding table replacement is not yet necessary, then significant material losses will incur. In **Figure 2**, two grinding tables are shown. On the left figure is a new grinding table, immediately after replacement, and on the right figure is a worn grinding table, straight before replacement.

Figure 2. New grinding table (left) and worn grinding table (right).

In practice which is common on plant A1, at thermoelectric power plant 'TEKO', Kostolac, grinding tables are replaced every 1800 h. However, it often happens that because of the increased presence of limestone, sand and other impurities in coal, grinding tables become deteriorated already after 1400 h, or even shorter. In that case, weaker effectiveness of the mill is noticeable, it is 'chocked', and serious problem with regulation occurs in an attempt to regulate the temperature of air mixture and pressure of fresh steam in front of the turbine. This appearance has for consequence significant misbalance of temperature distribution inside the firebox, which has negative influence on increased water injection in fresh steam, knockdown of coefficient of boiler efficiency and so on. In such conditions, usually, mill must be stopped unplanned for grinding table replacement and that incurs financial losses. Because of that, system which offers predictive maintenance is of great importance.

4. Proposed new predictive maintenance technique

The proposed solution to described problem is based on predictive maintenance. In this research, for the last step in predictive maintenance, the condition-monitoring data approach is chosen. This approach can be divided into two main categories: *model-based* prognostic technique and *data-driven* prognostic technique. Here, data-driven technique is chosen, because condition-monitoring data were available. Model-based method requires an accurate model of the system, which is highly complex. Maffezoni presents a useful physical model of the mill, the so-called *mass-balance model* with 76 *ordinary differential equations* (*ODE*), better known as a *knowledge-based model* [21]. It is obvious that it is very hard to make accurate model of the system, so this approach was not considered. On the other hand, the experience-based prognostic approach could not be used, because of the variable data statistics and an insufficient amount of data. For all these reasons, the data-driven approach was selected.

As it is described earlier, the first step in predictive maintenance programme is data acquisition. In this research, acoustic signals recorded in the vicinity of the mill were used to detect the condition of the mill. The acoustic signals were acquired from a coal mill at the 'TEKO' thermoelectric power plant, while it was operational. The main mill rotation frequency was about 12.5 Hz and the mill from which the signals were acquired had 10 impact plates.

Namely, in the literature it can be found that failure information is hidden in the spectral characteristics of vibration signals [23], but it has been demonstrated that in some cases acoustic signals are equally informative. In 2001, Baydar conducted a parallel analysis of the frequency characteristics of vibration signals and acoustic signals to detect various types of failures of rotary components, concluding that both signals can be used equally effectively [24]. The present research uses acoustic signals because they are simpler and less costly to record than vibration signals. They can also be acquired without interfering with mill operation because they are recorded externally.

The acoustic signals were acquired by means of a directional microphone at a distance of several millimetres, while the coal-grinding subsystem was operational. Recording of these signals is performed at the low altitude in thermoelectric power plant, where acoustic contamination is highly expressed. Because of that, special system for microphone fixation is projected, at a distance of several millimetres from the walls of analysed mill, so the power of useful signal could be multiple higher than the power of contaminating acoustic sources as neighbouring mills, feed pumps, surrounding valves and so on. The sampling frequency of recorded acoustic signals was 48 kHz. Data acquisition was conducted every 2 weeks on average, and it lasted for several minutes. **Table 1** shows the dates of recording, the dates of grinding table replacement and the duration of each signal. For faster implementation of the algorithm, the sampling frequency was decimated from 48 to 4.8 kHz, and the duration of the analysed signals was 1 min.

We can see from **Table 1** that the whole time period from the moment of grinding table replacement until the moment when grinding tables are worn is covered. After the first cycle of acoustic signal recording, three more recordings were performed after grinding table replacement. In this way, based on recorded acoustic signals, coal-grinding subsystem data are collected in different states. A large base of condition-monitoring data is obtained (without disturbing coal-grinding subsystem while it is operational) which can be further processed.

The second step in predictive maintenance is data processing. Given that collected data are acoustic signals, they are classified as waveform type of data. In order to overcome disadvantages encountered when such data are analysed in time domain and frequency domain [25], these data are analysed in time-frequency domain. A spectrogram was used to assess the acoustic signals in

Date of acquisition	Signal duration	Time since last maintenance
2 February 2012	10 min 51 s	14 days
24 February 2012	8 min 8 s	36 days
1 March 2012	8 min 8 s	42 days
15 March 2012	7 min 3 s	54 days
30 March 2012	6 min	6 days
5 April 2012	5 min	12 days
19 April 2012	6 min	26 days

Table 1. Recorded acoustic signals.

the time-frequency domain, which represented the spectral components of the signals in three dimensions very well: time information along the horizontal axis, frequency information along the vertical axis and amplitude depicted by a colour-coded scale. Colour intensity illustrated the strength of the spectral components. **Figure 3** [26] shows the spectrogram of an acoustic signal recorded on 30 March 2012, 6 days after grinding table replacement.

Figure 3 clearly shows the dominant frequencies, and indicates that they are the high harmonics of the basic frequency of mill rotation, which was f_0 =12.5 Hz. Also, the dominant peaks in the spectrum occurred at frequencies $10f_0$,$20f_0$ and so on, according to the fact that there were 10 impact plates inside the mill, such that the basic frequency of grinding table travelling alongside the microphone was $10f_0$. Given that the microphone was positioned so as to be as close as possible to the grinding table, these spectral components were much more pronounced than the other components.

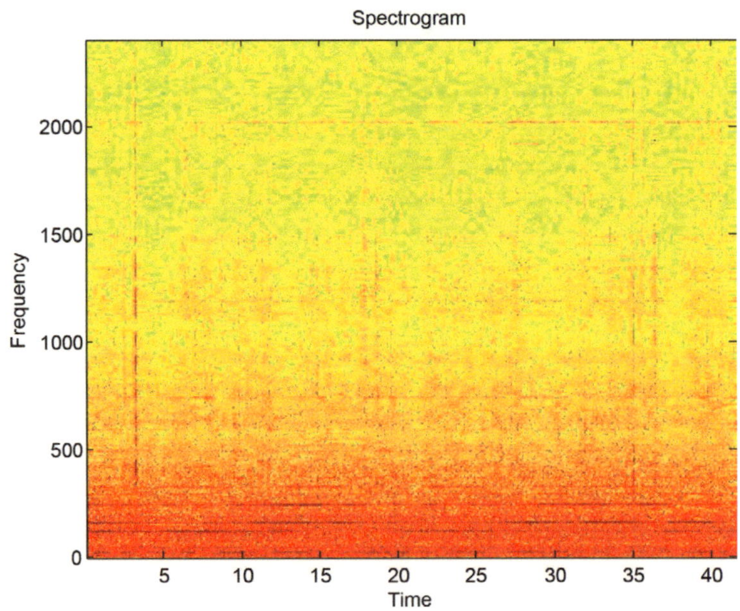

Figure 3. Spectrogram of acoustic signal.

After data acquisition, it was necessary to extract proper characteristics of the recorded acoustic signals in the frequency domain, in order to obtain vector of observations for analysis with T^2 control charts. As it was mentioned earlier, a spectrogram was used for acoustic signal representation. If recorded acoustic signal is denoted as $y[n]$, the spectrogram of acoustic signal S_p is often denoted as *short time fast Fourier transform* (*STFFT*) in literature [27] and computed as *fast Fourier transform* (*FFT*) on sliding window data. The idea of *STFFT* is dividing of the whole signal on segments with short time window, and applying the *Fourier transform* on each segment. The spectrogram represents a function of time and frequency arguments, which can be written as follows:

$$S_p = STFFT\{y[n]\} = S_p[f, n] \tag{1}$$

where f denotes the frequency and n the time argument of spectrogram.

The extracted quality characteristics in the frequency domain are the values of S_p across the time at the frequencies which represents the values around the high harmonics or the high harmonics themselves. Fourteen selected frequencies are shown in the vector f_p:

$$f_p = [14 \ \ 18.7 \ \ 23.4 \ \ 28.1 \ \ 32.8 \ \ 60.93 \ \ 126.5 \ \ 178.1 \ \ 187.5 \ \ 262.5 \ \ 346.8 \ \ 754.6 \ \ 1200 \ 2025] \tag{2}$$

Accordingly, the 14-dimensional vector of observations is formed at each time point:

$$X[n] = [x_1[n] \quad x_2[n] \quad \ldots \quad x_{14}[n]]^T \tag{3}$$

Coordinates of the vector $X[n]$ are calculated as follows:

$$x_i[n] = \sum_{j=n-L_w}^{n} S_p[f_i, j] \tag{4}$$

where f_i represents the ith coordinate of the frequency vector f_p, and L_w is the length of the window function. This is a procedure for the generation of the initial observation vector. In this way, the data-processing step and feature extraction are completed.

The last step in predictive maintenance programme is maintenance decision-making. As it is described in the beginning of this section, data-driven technique is chosen, that is, it is decided that the input of the sequence of observations be analysed with T^2 control charts, and then, outputs of control charts will be the input sequence for *hidden Markov model* (*HMM*). *HMM* should give us the information about the grinding tables condition, that is, are they worn so that their replacement is necessary. This would be the second approach in failure prognostic, because of the prediction that the system will work without failure until some future time, that is, until the next interval when inspection is needed.

After obtaining the vector of observations, T^2 control charts were constructed. Generally speaking, a control chart is a statistical tool used to detect failure. Control charts make a clear distinction between common causes of variations in the process and failures of the system. For a system where only common causes of variations are present, we say that such a system is under *statistical control*. A control chart generally has a centre line (*CL*), upper control limit (*UCL*) and lower control limit (*LCL*). The centre line represents the mean value of the quality characteristic of interest, detected while the process is under statistical control. The control limits are selected such that while the process is under statistical control, nearly all the points in the control chart will fall between these two lines.

The first step in constructing the control charts requires an analysis of preliminary data, which are under statistical control. This step is called Phase I, and data used in this phase are called the historical data set. In Phase II, the control chart is used to monitor the process by comparing the sample statistic for each successive sample as it is drawn from the process to the control limits established in Phase I [28, 29].

A multivariate analysis with Hotelling T^2 control charts was undertaken in the present research [30]. Based on observation vectors, T^2 sequence of values may be calculated according to the following equation:

$$T^2[n] = (X[n] - \overline{X})^T S^{-1}(X[n] - \overline{X}) \tag{5}$$

where $\overline{X}$ and S denote the sample estimators of the mean value vector and the covariance matrix, respectively. Assuming that during the data acquisition sequence of N observations $\{X[0], X[1], \ldots, X[N-1]\}$ is generated, sample estimators of vector of mean values and covariance matrix can be written as follows:

$$\overline{X} = \frac{1}{N}\sum_{i=0}^{N-1} X[i] \tag{6}$$

$$S = \frac{1}{N-1}\sum_{i=0}^{N-1}(X[i] - \overline{X})(X[i] - \overline{X})^T \tag{7}$$

The control limits in Phase II are

$$UCL = \chi^2_{(\alpha,p)}, \; LCL = 0 \tag{8}$$

where $\chi^2_{(\alpha,p)}$ is the upper α percentage point of the chi-squared distribution with p degrees of freedom (p represents the number of variables which is in our case 14). When the number of preliminary samples n is large ($n > 100$), using chi-squared control limit in Phase II is reasonable approximation [29]. In Phase I, the limits are based on beta distribution:

$$UCL = \left[\frac{(n-1)^2}{n}\right]\beta_{(\alpha;p/2,(n-p-1)/2)}, \; LCL = 0 \tag{9}$$

where $\beta_{(\alpha;p/2,(n-p-1)/2)}$ is the upper α percentile of beta distribution with parameters $p/2$ and $(n-p-1)/2$.

According to relation (5), the time sequence of T^2 values is formed, denoted as $\{T^2[0], T^2[1], \ldots, T^2[n]\}$ where n denotes the sequence number of sliding window data. In order to account for system dynamics, instead of the very last control chart sample, the last 10 samples were used for the characterization of the actual state of grinding tables. In other words, vector

$$O[n] = [\,T^2[n-9] \quad T^2[n-8] \quad \ldots \quad T^2[n]\,]^T \tag{10}$$

will be used for further estimation of system states. However, if this vector had been introduced as observation in *HMM*, it would be necessary to estimate joint probability function for this, tenth-dimensional vector. In order to avoid this complex numerical problem, it has been decided, as it is usual in the literature, to apply the procedure of *vector quantization*. In this purpose, the method of *k-means* clustering is used [31]. The result of *k-means* clustering is the sequence of *k*-cluster centres (centroids). In our case, based on *try-and-error* approach, it turned out that for $k = 4$ satisfying results are gain and cluster centres $(C_i, i = 1, 2, 3, 4)$ are obtained. Accordingly, the final vectors of observations $\hat{O}[n]$ are formed and forwarded to *HMM* in the following way:

$$\min_j \|O[n] - C_j\|^2 = \|O[n] - C_k\| \Rightarrow \hat{O}[n] = C_k \tag{11}$$

After the samples were coded as described above, the next step was to construct the *HMM*. An *HMM* is a statistical model used to describe the transition of a system between states. It is an extension of the ordinary Markov chains with non-observable or partially observable states. Generally, *HMM* has N states $S = \{S_1, S_2, \ldots, S_N\}$ and M observation symbols $V = \{v_1, v_2, \ldots, v_M\}$. *HMM* with three states is shown in **Figure 4**. The states are connected in such a way that it is possible to move from any one to the other. The hidden state at time t is denoted by q_t, and the

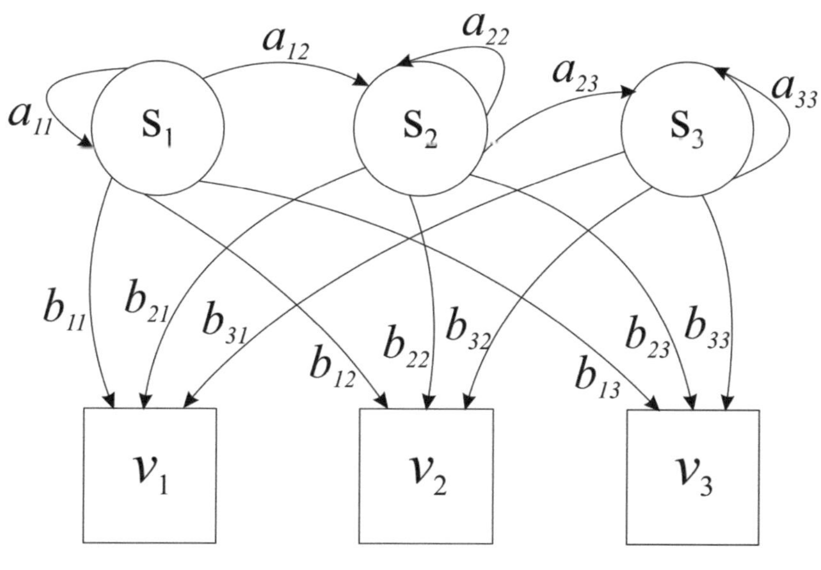

Figure 4. *HMM* with three states.

move from one state to the other is subject to Markov's rule (that state q_t depends solely on state q_{t-1}). In addition to the number of states, *N*, and the number of observation symbols, *M*, several other *HMM* characteristics need to be defined.

The transition matrix $\mathbf{A} = \{a_{ij}\}$ represents the probability of moving from state *i* to state *j*. The coefficients a_{ij} are non-negative in the general case, and equal to zero if there is no direct switching from one state to another. The sum of probabilities in each matrix of type **A** needs to be equal to 1. The observation matrix (also called the emission matrix) $\mathbf{B} = \{b_j(k)\}$ shows the probability that observation *k* was produced by the *j*th state.

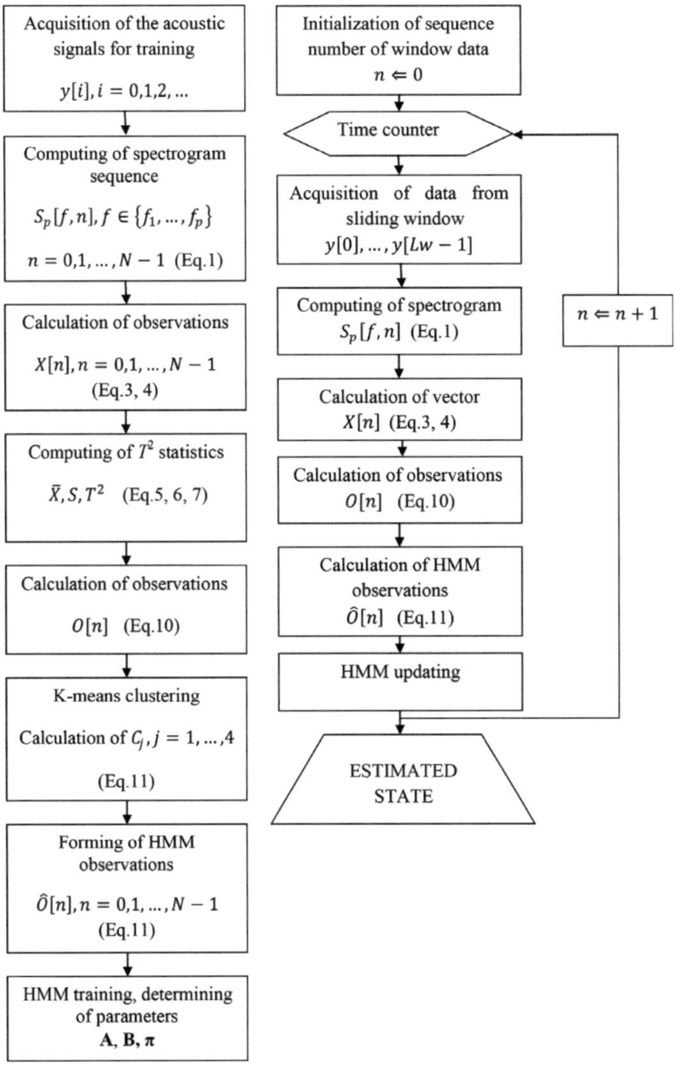

Figure 5. Flow diagram of the proposed algorithm: *offline* procedure (left) and *online* procedure (right).

The sequence of initial states $\pi = \{\pi_i\}$ carries information about initial probabilities, indicating the likelihood that a new input sequence will move from a given state. Finally, the *HMM* can be defined by the triplet:

$$\lambda = (\mathbf{A}, \mathbf{B}, \pi) \tag{12}$$

There are three fundamental problems that can be solved by means of *HMM*. A detailed description of *HMM*s and the solutions to these three problems is available in Ref. [32].

Figure 5 [26] shows how the proposed algorithm for predictive maintenance is organized. For the purpose of the practical implementation of the proposed method, it should be clarified that certain activities are realized only once (like *offline* procedure) in order to determine the necessary statistics and *HMM* training. On the other hand, once the *offline* procedure is over, the algorithm can be implemented in real time and thus providing *online* monitoring of the mill-grinding plates states.

5. Results

In this chapter, gained results after applying the proposed technique for predictive maintenance on described problem at thermoelectric power plant will be presented. As it is previously explained, after data acquisition and feature extraction from recorded acoustic signals, T^2 control charts are formed.

The acoustic signal recorded on 30 March 2012 was used for $\overline{X}$ and S estimation in Eqs. (6) and (7), knowing that a new grinding table was operational. In this way, this signal was observed as historical data set. This was in effect Phase I of statistical control, where the entire coal-grinding subsystem was under statistical control. The estimated values of $\overline{X}$ and S in Phase I were to be used in Phase II of the multivariate analysis. The chi-squared control limit was taken as the *UCL*, as in Eq. (8). For the 14 quality characteristics, $UCL = 36.12$ (for the value $\alpha = {}_{0.001}$) and $LCL = 0$. In order to justify the using of chi-squared control limit, in **Figure 6**, *Q-Q* plot [29] with T^2 quantiles on y-axis and chi-squared quantiles on x-axis are shown. For illustration, *Q-Q* plot for T^2 values for signals recorded on 30 March 2012 is shown, that is, for the signal recorded 6 days after grinding table replacement. During research, this check is done for all the signals in order to confirm that the choice of chi-squared control limit is justified.

From **Figure 6**, we can see that the values follow chi-squared distribution, that is, the figure shows approximately linear trend along the line of 45°, except the last few points which are slightly away from the projected trend line. Before T^2 control charts were constructed, we expected that the number of outliers will increase as grinding tables become worn out. **Figure 7** [26] shows the T^2 control chart for the acoustic signal recorded on 2 February 2012, 2 weeks after grinding table replacement.

Figure 8 [26] shows the T^2 multivariate control chart for the acoustic signal recorded on 24 February 2012, 5 weeks after grinding table replacement.

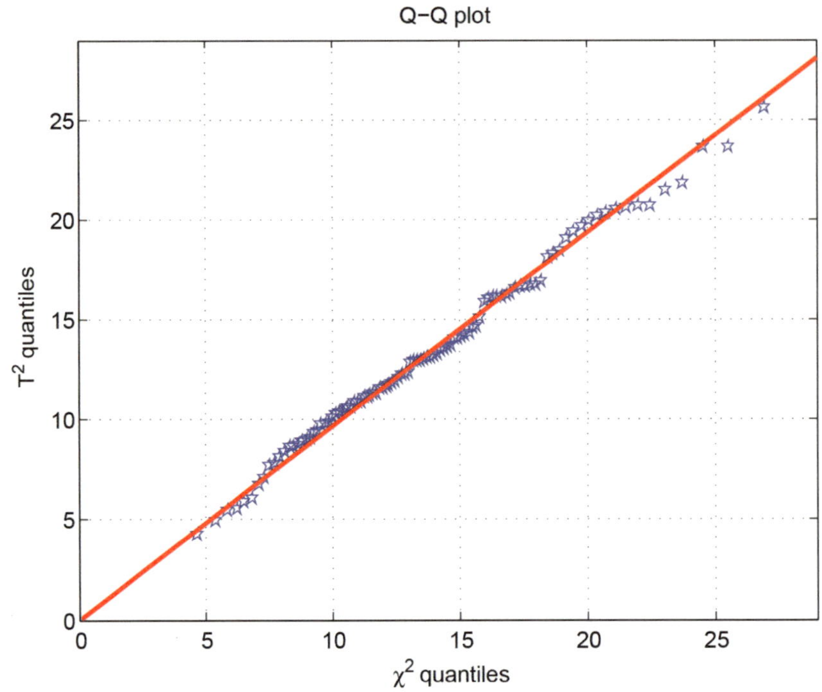

Figure 6. *Q-Q* plot for recorded acoustic signal 6 days after grinding table replacement.

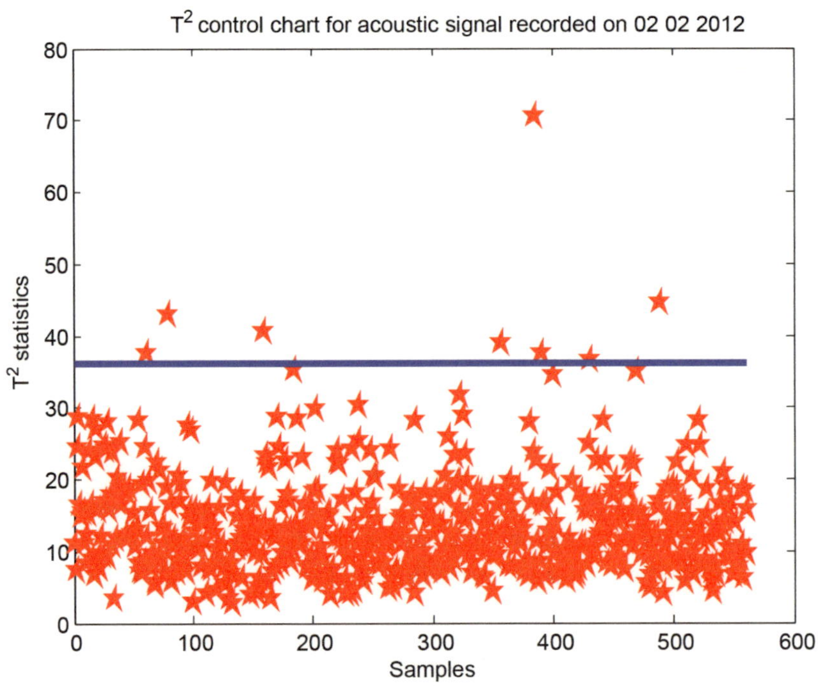

Figure 7. T^2 control chart for acoustic signal recorded 2 weeks after grinding table replacement.

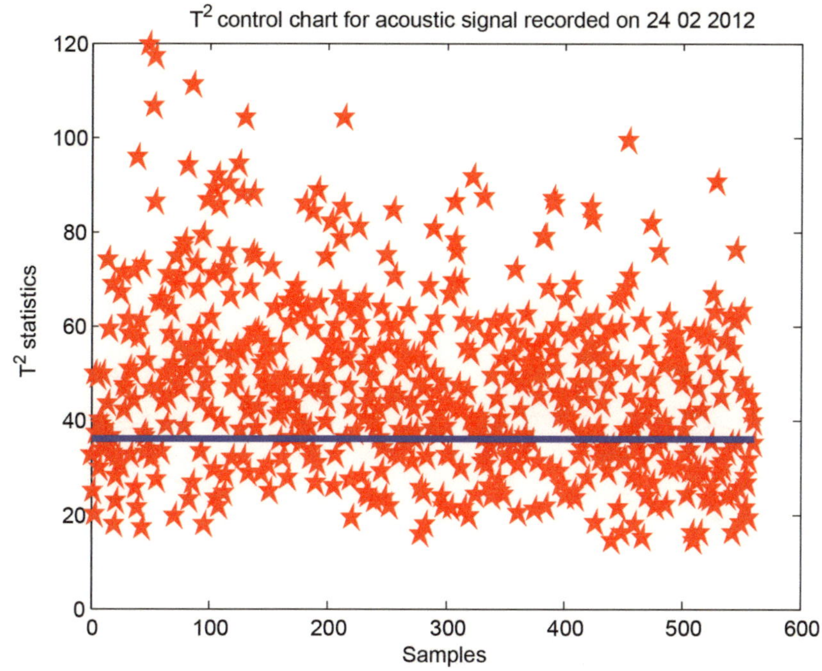

Figure 8. T^2 control chart for acoustic signal recorded 5 weeks after grinding table replacement.

Figure 9 [26] shows the T^2 control chart for the acoustic signal recorded on 15 March 2012, 8 weeks after grinding table replacement.

It is apparent from **Figures 7–9** that the number of points above the *UCL* on the T^2 control chart grew as the grinding table became increasingly worn. Eight weeks after replacement, nearly all the points were beyond the *UCL*. To confirm the results, the multivariate analysis was repeated using the signals recorded on 5 and 19 April 2012. Table 2 shows the exact number of outliers for all the recorded signals for the different values of *UCL* (i.e. for different values of parameter α).

The difference in the number of points above the *UCL* for the signals recorded on 2 February and 19 April 2012 can be explained. Namely, both signals were acquired 2 weeks after grinding table replacement, but the results are different for two reasons: (1) The signal acquisition conditions were not ideal because of noise. All the recorded signals reflect this noise, as well as other disturbances (e.g. when a large chunk of coal or stone hits the mill). The signals were not filtered, because of the possible information loss. All this could have influenced the accuracy of the results. (2) Grinding table wear depends on the quality of the coal and of the grinding table itself. It is therefore impossible to ascertain what the right time for grinding table replacement would be, unless the entire subsystem is shut down and opened for visual inspection.

According to **Table 2**, we can conclude that with the choice of parameter α = 0.001, 'over controlling' control chart is constructed, while with the choice of parameter α = 0.025, false alarm rate is too large. Anyway, no matter which value of *UCL* we have chosen, the number of outliers is larger as grinding tables are getting worn out. Namely, in the proposed method

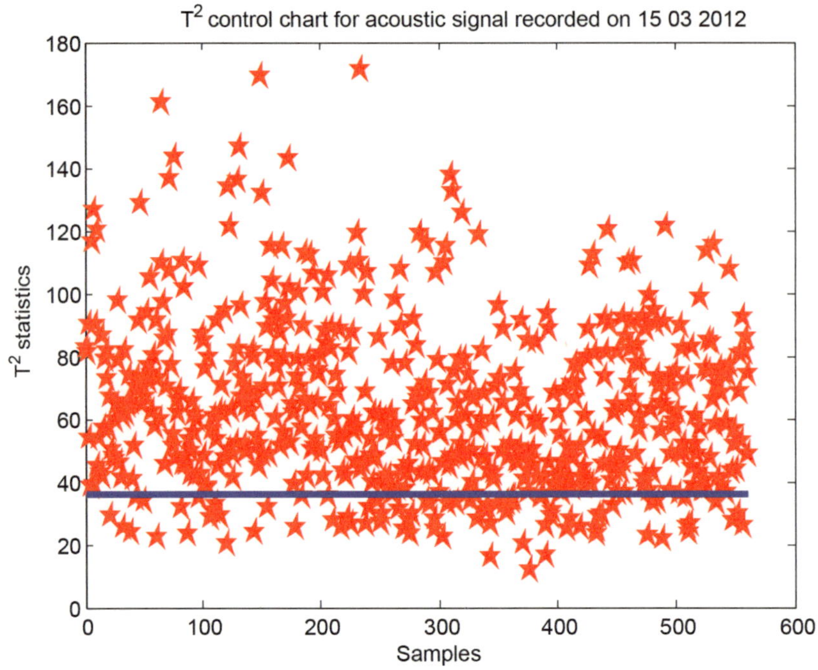

Figure 9. T^2 control chart for acoustic signal recorded 8 weeks after grinding table replacement.

control charts were not used for classical fault detection, yet for forming of T^2 statistics that will be parameterized for making the *HMM* observations. The choice of the *UCL* does not have an influence on T^2 statistics value, that is, on forming of observations for *HMM*. Thus, the choice of parameter α, that is, making of compromise between the first type error and the second type error, does not have an influence on observation values for *HMM*, which is not usually the case

Date of recording	Number of weeks after grinding table replacement	Number of points above UCL (%), α = 0.001, UCL = 36.12	Number of points above UCL (%), α = 0.005, UCL = 31.32	Number of points above UCL (%), α = 0.01, UCL = 29.14	Number of points above UCL (%), α = 0.025, UCL = 26.12
2 February 2012	2 weeks	1.43%	2.14%	2.46%	5%
24 February 2012	5 weeks	68.27%	79.5%	83.78%	88.41%
15 March 2012	8 weeks	84.85%	90.91%	92.87%	95.54%
05 April 2012	2 weeks	16.75%	27.63%	32.98%	43.14%
19 April 2012	4 weeks	57.58%	70.05%	74.87%	81.64%

Table 2. Number of points above UCL.

when classical control chart needs to detect the fault and when the choice of parameter α has large influence for the correct determination of *UCL*.

After T^2 control charts were constructed, vector quantization was undertaken, as described in the previous section, in order to represent the control chart samples as a sequence of observations for the *HMM*. **Figure 10** [26] shows the estimated probability density functions of the T^2 control chart samples for the signals recorded 2, 5 and 8 weeks after grinding table replacement. It is apparent that the T^2 statistics change over time and that they are a function of the condition of the grinding table (i.e. they change as the condition of the grinding table changes).

The final step of the proposed algorithm was to construct the *HMM*. The states of *HMM* are chosen so to represent the physical condition of mill-grinding plates. In order to illustrate the proposed method, it is assumed that *HMM* has three states. The first state is the condition of the grinding table immediately after replacement (i.e. that of a new grinding table). Having in mind that the average length of mill-grinding table duration is 1600 h approximately, the fact that *HMM* is in the first state could be interpreted as the grinding tables being in the first third of their life. The second state was the 'intermediate state', where the grinding table becomes partially worn out, but there is still time before replacement is needed. Consequently, the system staying in the second state can be interpreted as the grinding tables entering the second third of their lifetime. The third state means that the condition of the grinding table had deteriorated to the point where replacement is necessary. Namely, this research started from the assumption that *HMM* has only three states, but if it is needed that the grinding table

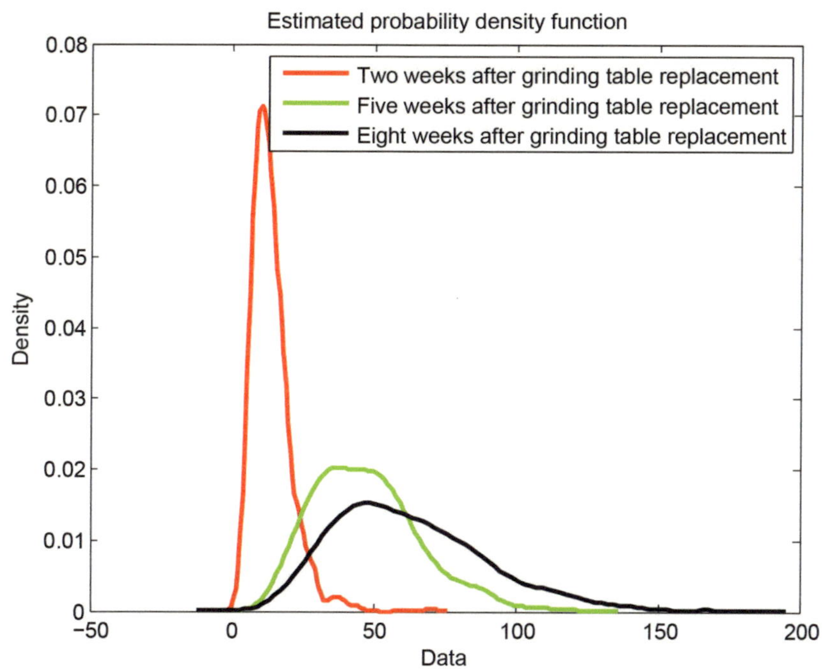

Figure 10. Estimated probability density functions for signals recorded 2, 5 and 8 weeks after grinding table replacement.

conditions are characterized with greater precision, the number of states could be increased. **Figure 11** [26] shows the sequence of observations and corresponding HMM states.

It is apparent from **Figure 11** that the *HMM* provides information about a change in the condition of the grinding table. It is obvious that the time of *HMM* entry into the third state (worn-out grinding table) coincides with the beginning of observations that correspond to the control chart samples for the signal recorded 8 weeks after replacement.

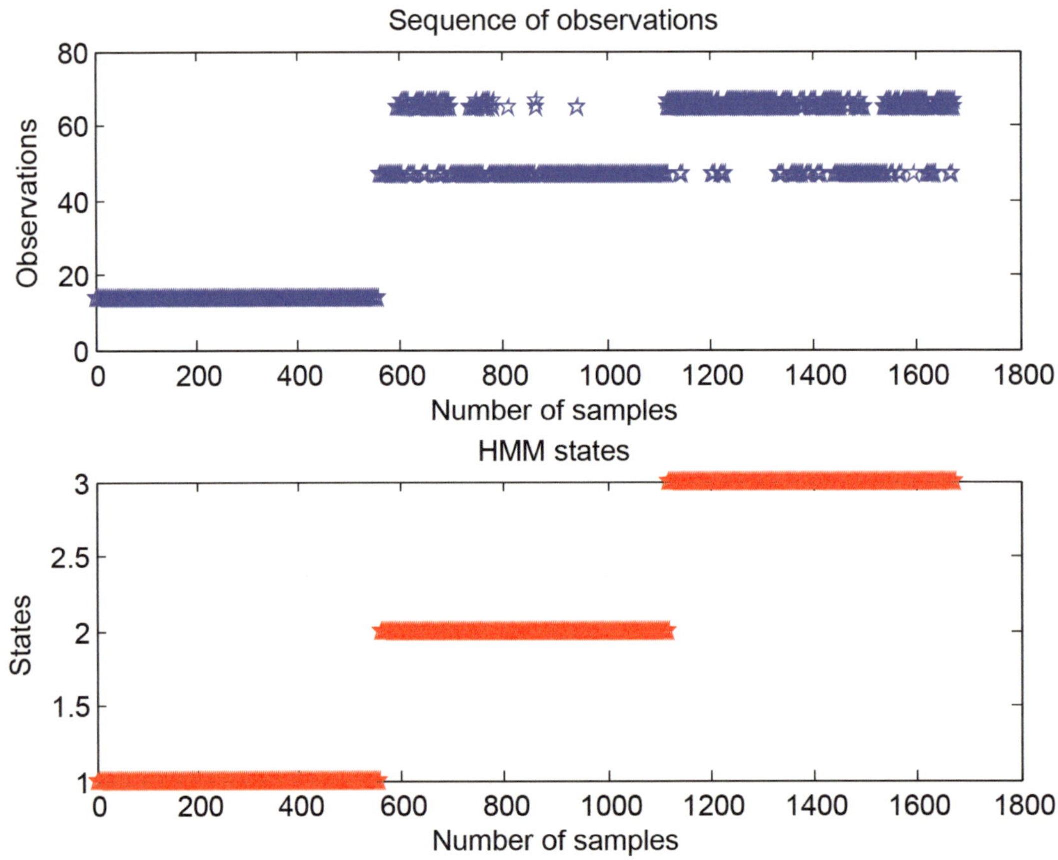

Figure 11. *HMM* states.

6. Conclusion

Based on the presented results, we can make several conclusions. Firstly, the assumption set at the beginning of this research, that useful information from spectral components of acoustic signals can be extracted is confirmed. Based on this information, the condition of rotating elements of the mill can be recognized. As it is previously explained, in the literature there are mostly preferred vibration signals in regard to the acoustic signals, when we talk about informative content. Given that the recording of acoustic signals is much cheaper than the recording of vibration signals, and processing of acoustic signals is much simpler from vibration signals processing, confirmation about informative content of acoustic signals is very important.

The originality of the proposed method is a combination of control charts and *HMMs* in failure prognostic, as well as in the application of control charts on extracted components from spectrogram. Namely, in the literature one can find control charts whose construction is based on spectral analysis of the signal [33]. Here, a different approach is proposed, that is, to apply the T^2 control charts on spectral components of the signal. Based on the results, this approach has proven to be very efficient. In the literature, one can also find the application of control charts and *HMMs* for degradation process diagnosis [34], as well as for fault detection [35], but in these papers standard *p*-charts and Hotelling T^2 control charts are used. Reports of other research dealing with the detection of certain types of failures at thermoelectric power plants can be found in the literature [36–38]. Also, *HMM*-based diagnostic models founded upon the condition of the system can be found in Refs. [39, 40]. In regard to all mentioned references, the original approach is proposed here.

As it is previously described, in the case of failure prognostic, in literature the most common approach is the first approach, that is, the estimation when the fault will occur (*RUL* estimation). In this research, the accent is on the second approach, that is, on the estimation of probability that the machine will work without failure until some future time (in our case, until the next interval when inspection and grinding table replacement are needed). With the proposed method, *HMM* gives us the information about grinding tables condition, that is, when the grinding tables are worn out, so that their replacement is needed.

The advantage of the proposed method is that it is non-invasive, because for the acquisition of condition-monitoring data it is not necessary to interrupt coal-grinding subsystem operation and shut down the whole subsystem. Another advantage is that it is based on acoustic signals processing which are simpler for processing and acquisition in regard to vibration signals. Software realization of the proposed algorithm is not too much complex and it is not time consuming when *HMM* is once trained.

A shortcoming of this method is the recording of acoustic signals in the presence of the unavoidable noise, which can influence on the accuracy of the results. Presented results are gathered *offline*, that is, *HMM* is trained based on the already recorded signals. For applying this method on *online* data, much larger amount of data are needed for adequate *HMM* training and more accurate determining of time moment when the grinding table replacement is needed. Anyhow, the proposed method can be applied in real time and used for higher stability and reliability of one of the most important subsystems in a thermoelectric power plant.

Further direction in this research would be the making of an adaptive system which would be adjustable to new statistics which are consequences of components ageing, not just the condition of grinding tables plates. Also, significant study could be made when condition-monitoring data would be recorded vibration signals, for comparative analysis with acoustic signals. Additional event data could upgrade the proposed method in combination with condition maintenance data. Some future research could be to make optimal maintenance policy in thermoelectric power plant, according to gain results.

Author details

Emilija Kisić[1]*, Željko Đurović[2] and Vera Petrović[1]

*Address all correspondence to: emilija.kisic@viser.edu.rs

1 School of Electrical Engineering and Computer Science of Applied Studies, Belgrade, Serbia

2 School of Electrical Engineering, University of Belgrade, Belgrade, Serbia

References

[1] Rosmaini A, Shahrul K. An overview of time-based and condition-based maintenance in industrial application. Journal of Computers & Industrial Engineering. 2012;**63**:135–149. DOI: 10.1016/j.cie.2012.02.002

[2] Bloch H, Geitner F. Machinery Failure Analysis and Trouble Shooting. USA: Gulf Professional Publishing; 1997

[3] Mobley R. An Introduction to Predictive Maintenance. USA: Elsevier Science; 2002

[4] Wikipedia. List of Industrial Disasters [Internet]. 2017. Available from: https://en.wikipedia.org/wiki/List_of_industrial_disasters#cite_note-18 [Accessed: 20 February 2016]

[5] Jardine A, Lin D, Banjevic D. A review on machinery diagnostics and prognostics implementing condition-based maintenance. Mechanical Systems and Signal Processing. 2006;**20**:483–1510. DOI: 10.1016/j.ymssp.2005.09.012

[6] Vachtsevanos G, Lewis F, Roemer M, Hess A, Wu B. Intelligent Fault Diagnosis and Prognosis for Engineering Systems. NJ: Wiley; 2006

[7] Ding S. Model-Based Fault Diagnosis Techniques. Berlin: Springer; 2008

[8] Heng A, Zhang S, Tan A, Mathew J. Rotating machinery prognostics: State of the art, challenges and opportunities. Mechanical Systems and Signal Processing. 2009;**23**:724–739. DOI: 10.1016/j.ymssp.2008.06.009

[9] Ma L. Condition monitoring in engineering asset management. In: Proceedings of Asia-Pacific Vibration Conference; 6–9 August 2007; Sapporo, Japan: APVC2007; 2007. pp. 1–16

[10] Le Son K, Fouladirad M, Barros A, Levrat E, Iung B. Remaining useful life estimation based on stochastic deterioration models: A comparative study. Reliability Engineering and System Safety. 2013;**112**:165–175. DOI: 10.1016/j.ress.2012.11.022

[11] Sankararaman S. Significance, interpretation, and quantification of uncertainty in prognostics and remaining useful life prediction. Mechanical Systems and Signal Processing. 2015;**52–53**:228–247. DOI: 10.1016/j.ymssp.2014.05.029

[12] Asadzadeh S, Azadeh A. An integrated systemic model for optimization of condition based maintenance with human error. Reliability Engineering and System Safety. 2014;**124**:117–131. DOI: 10.1016/j.ress.2013.11.008

[13] Banjevic D, Jardine A. Calculation of reliability function and remaining useful life for a Markov failure time process. IMA Journal of Management Mathematics. 2006;**17**:115–130. DOI: 10.1093/imaman/dpi029

[14] Kacprzynski J, Sarlashkar A, Roemer M. Predicting remaining life by fusing the physics of failure modeling with diagnostics. Journal of Metal. 2004;**56**:29–35. DOI: 10.1007/s11837-004-0029-2

[15] Lin A, Makis V. Recursive filters for a partially observable system subject to random failure. Advances in Applied Probability. 2003;**35**:207–227. DOI: 10.1017/S0001867800012155

[16] Groer P. Analysis of time-to-failure with a Weibull model. In: Proceedings of the Maintenance and Reliability Conference; Knoxville, TN, USA; 2000. pp. 59.01-59.04

[17] Schwabacher M. A survey of data-driven prognostics. In: Proceedings of the AIAA Infotech@Aerospace Conference; 2005; Arlington, Virginia: American Institute of Aeronautics and Astronautics; 2005. pp. 1–5

[18] Luo J, Pattipati K, Qiao L, Chigusa S. Model-based prognostic techniques applied to a suspension system. Transactions on Systems, Man, and Cybernetics. 2003;**38**:1156–1168. DOI: 10.1109/TSMCA.2008.2001055

[19] Moubray J. Reliability-Centred Maintenance. Oxford: Butterworth-Heinemann; 1997

[20] Tobon-Mejia D, Medjaher K, Zerhouni N, Tripot G. Hidden Markov models for failure diagnostic and prognostic. In: Prognostics and Health Management Conference; 2011; Shenzhen, China. IEEE Catalog Number: CFPII6IH-PRT; 2011. pp. 1–8

[21] Flynn D. Thermal Power Plant Simulation and Control. London: The Institution of Electrical Engineers; 2003

[22] Pradeebha P, Pappa N. Modelling and outlet temperature control of coal mill using model predictive controller. IEEE International Conference on Control Applications (CCA); Hyderabad, India; 2013. pp. 1069–1074. DOI: 10.1109/CCA.2013.6662893

[23] Renwick J, Babson P. Vibration analysis – a proven technique as a predictive maintenance tool. IEEE Transaction on Industry Applications. 1985;**IA-21**:324–33.

[24] Baydar N, Ball A. A comparative study of acoustic and vibration signals in detection of gear failures using Wigner-Ville distribution. Mechanical Systems and Signal Processing. 2001;**15**:1091–1107. DOI: 10.1006/mssp.2000.1338

[25] Cohen L. Time-frequency distribution – A review. Proceedings of the IEEE. 1989;**77**:941–981

[26] Kisić E, Đurović Ž, Kovačević B, Petrović V. Application of T^2 control charts and hidden Markov models in condition-based maintenance at thermoelectric power plants. Shock and Vibration. 2015;**2015**:1–11. DOI: 10.1155/2015/960349

[27] Andrade A, Esat I, Badi M. Gearbox fault detection using statistical methods, time-frequency methods (STFT and Wigner-Ville distribution) and harmonic wavelet-A comparative study. In: Proceedings of the COMADEM '99; UK: Chipping Norton; 1999. pp. 77–85

[28] Montgomery D. Introduction to Statistical Quality Control. 5th ed. New York, NY: John Wiley & Sons; 2005

[29] Mason R, Young J. Multivariate Statistical Process Control With Industrial Applications. USA: ASA-SIAM; 2002

[30] Hotelling A. Multivariate quality control. In: Eisenhart C, Hastay M, Wallis W, editors. Techniques of Statistical Analysis. New York, NY: McGraw-Hill; 1947. pp. 111–184

[31] Duda R, Hart P, Stork D. Pattern Classification. New York, NY: John Wiley & Sons; 2001

[32] Rabiner L. A tutorial on hidden Markov models and selected applications in speech recognition. Proceedings of the IEEE. 1989;**77**:257–286

[33] Tiplica T, Kobi A, Barreau A, Chassagnon R. Spectral control chart. Quality Engineering. 2005;**17**:695–702. DOI: 10.1080/08982110500251287

[34] Lee S, Li L, Ni J. Online degradation assessment and adaptive fault detection using modified hidden Markov model. Journal of Manufacturing Science and Engineering. 2010;**132**:0210101-02101011. DOI: 10.1115/1.4001247

[35] Tai A, Ching W, Chan L. Detection of machine failure: Hidden Markov model approach. Computers & Industrial Engineering. 2009;**57**:608–619. DOI: 10.1016/j.cie.2008.09.028

[36] Chen K, Chen L, Chen M, Lee C. Using SVM based method for equipment fault detection in a thermal power plant. Computers in Industry. 2011;**62**:42–50. DOI: 10.1016/j.compind.2010.05.013

[37] Yang S. A condition-based preventive maintenance arrangement for thermal power plants. Electric Power Systems Research. 2004;**72**:49–62. DOI: 10.1016/j.epsr.2004.03.007

[38] Frank P, Ding S, Marcu T. Model-based fault diagnosis in technical processes. Transactions of the Institute of Measurement and Control. 2000;**22**:57–101. DOI: 10.1177/014233120002200104

[39] Wang L, Mehrabi M, Kannatey-Asibu E. Hidden Markov model-based tool wear monitoring in turning. ASME Journal of Manufacturing Science and Engineering. 2002;**124**:651–658. DOI: 10.1115/1.1475320

[40] Li Z, Wu Z, He Y, Fulei C. Hidden Markov model-based fault diagnostics method in speed-up and speed-down process for rotating machinery. Mechanical Systems and Signal Processing. 2005;**19**:329–339. DOI: 10.1016/j.ymssp.2004.01.001

Chapter 4

Model Development for Analysis of Steam Power Plant Reliability

Bernard Tonderayi Mangara

Additional information is available at the end of the chapter

http://dx.doi.org/10.5772/intechopen.68285

Abstract

The graph theoretical analysis and a graph's characteristic polynomial are deployed as the basis for a system approach employed to develop a model for estimating the reliability index and evaluating the availability index for a coal-fired generating power station. In this research, the coal-fired generating station system is divided into six sub-systems. Elementary to evaluating the reliability (estimate) of the said system is the consideration of all the sub-systems and their interrelations. Approximate reliability attributes of the graph are used to model the approximate reliability of the coal-fired generating station. Sub-system reliability is represented by the nodes in the graph, and the links represent the reliability of interrelations of these sub-systems. Computing a graph's characteristic polynomial using three different methods, namely, the linearly independent cycles, the figure equation and the adjacent matrix, the approximate reliability of the system is determined. Three methods are used, for comparison purposes, as estimating reliability is always an imperfect endeavour. The methodology proposed in this study is illustrated step-by-step with the help of examples.

Keywords: Reliability, coal-fired, steam, method, components

1. Introduction

Figure 1 shows the schematic diagram of a coal-fired generating station [1–3]. The energy conversion in a coal-fired generating station is as follows [1–3]. Coal is conveyed to a mill and the mill crushes the coal into fine powder, which is pulverized. Thereafter, the pulverized fuel is blown into the boiler where it mixes with a supply of pre-heated air for combustion. The said combustion of a mixture of pulverized fuel and pre-heated air in the boiler produces steam, at high temperatures and pressures, which is passed through the steam turbine. The boiler drives the steam turbine, which is coupled to the electricity generator. The generator then supplies the national electrical load. In spite of the advances in the design and materials in the last few

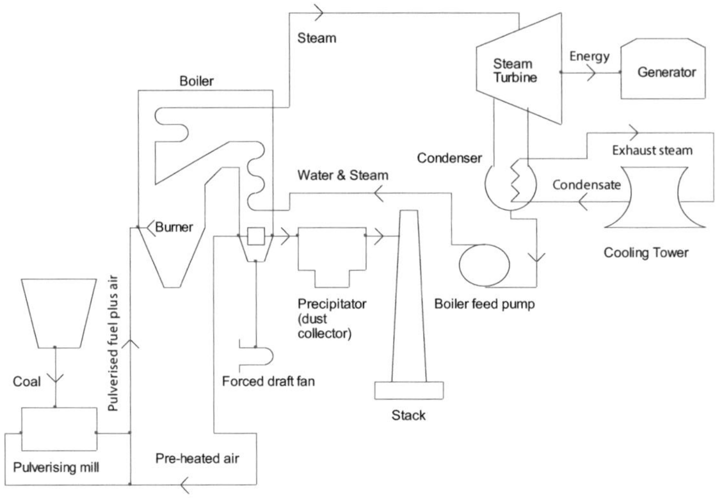

Figure 1. Schematic view of a coal-fired generating station.

years, the efficiency of the most modern coal stations is in the region of 40%. Therefore, the 60% of energy rejected as heat forms the exhaust steam from the low pressure turbine, which is cooled to form condensate by the passage through the condenser of large quantities of sea- or river-water. If the station is located inland or if there is concern over the environmental effects of raising the temperature of the sea- or river-water, cooling towers are then used.

Steam power-stations operate on the Rankine cycle [1]. The Rankine cycle, in a steam power-station, is modified to include super-heating, feed-water heating, and steam re-heating. To achieve high efficiency, the steam has to be used at maximum possible pressures and temperatures. Furthermore, for economic construction of turbines, the larger the size, the less the capital cost per unit of power output. Consequently, turbo-generator sets of 500 MW and more have been used. For steam turbines 100 MW and above, efficiency is increased by using an external heater to re-heat the steam, after it has been partially expanded. In addition, the re-heated steam is then returned to the turbine where it is expanded through the final stages of blading.

The study of reliability of complex systems such as steam power plants is of interest for power utility companies. This is so because the power utilities have to minimize operation and maintenance expenses while ensuring the reliability, safety, and security of supply to their national electrical load in order to remain competitive in the global market. In this study, complex systems are defined as large collections of interconnected components whose interactions lead to macroscopic behaviours [2, 3]. For complex systems, it is required to translate system reliability requirements into detailed specifications for all components that constitute the system. This process is often referred to as the reliability apportionment. During reliability apportionment, the reliability analyst has to perceive and develop the relationships between component, sub-system, and system reliabilities. The decisive role in this process is in understanding and quantifying the reliability importance of different parts of the equipment.

Steam power plant reliability encompasses a range of issues related to the design and analysis of these power plant generating networks. Furthermore, the said coal-fired thermal power plant components are prone to random failures. Generally, network models are comparatively simple, and yet quite generic. Varied applied problems of steam power plant environments can be modelled with networks [2, 3]. In the field of steam power plant reliability, the final goal of research is to provide design engineers with procedures to further improve the quality of designs whereupon reliability is a significant factor to take into account.

In this study, a system is defined as a bounded physical entity that achieves in its domain a defined objective through interaction of its components ([4], pp. 604). It follows from this definition of system that the following notation is going to be used for system reliability ([4], pp. 148):

Since the state variables $X_i(t)$ for $i = 1,\ 2, \ldots,\ n$ are binary, then

$$E[X_i(t)] = 0 \cdot \Pr(X_i(t) = 0) + 1 \cdot \Pr(X_i(t) = 1) = p_i(t), \quad \text{for } i = 1, 2, \ldots, n \tag{1}$$

Similarly the system reliability (at time t) is

$$p_s(t) = E\Big(\varnothing\big(X(t)\big)\Big) \tag{2}$$

It can be shown that when the components are independent, the system reliability, $p_s(t)$, will be a function of the $p_i(t)'s$ only. Hence, $p_s(t)$ may be written as follows

$$p_s(t) = h\big(p_1(t), p_2(t), \ldots, p_n(t)\big) = h\big(p(t)\big) \tag{3}$$

Unless stated otherwise, the letter h will be used to express system reliability in situations *where the components are independent*.

The size of the state space is one of the major impediments in steam power plant system reliability analysis. For a complex and large-scale power plant system, the number of system states is enormous. It can be noted that a system consisting of n components and each component with binary states (working or failed) has a total of 2^n states. For example, if one considers a case when n is 300, the number of states is 2.04×10^{90}. For the preceding example, if one would analyse all the possible states individually in order to identify the contingencies that help to bring about the system unreliability, this would require much computational effort. Furthermore, it would be impractical for typical steam power plants. Therefore, there is need to choose a methodology which reduces the state space, and a subsequent selection and evaluation of contingencies. The graph theoretical analysis (GTA) method [5, 6] is chosen for this study.

Graph theory is a branch of mathematics which has existed for many years not only as an area of mathematical study but also as a tool for intuition and illustration [7]. Graphs can be used in wiring diagrams to represent the physical elements of an electrical circuit; a street map is also a graph with the streets as links (edges), intersections of streets as nodes (vertices), and street names as labels of the links (edges). In the above-mentioned cases, the graphs resemble the physical objects that they represent. Thus, the application (and sometimes the genesis) of the graph-theoretic ideas is immediate. Computer program flow diagrams and road maps with one way streets are examples of graphs that contain the *concept of direction or flow to the links (edges)*; and these are called *directed graphs*. The applications of graphs and directed graphs in almost all areas of the physical sciences and mathematics have been known for half a century or more years [7]. In this study, graph-theoretic ideas are applied to some of the fundamental topics in power plant engineering. While there are many such applications, we shall focus on only using graph-theoretic ideas for estimating the reliability index and evaluating the availability index for a coal-fired generating power station.

Graph theory has been successfully used to model many different types of systems, inclusive of coal-based steam power plants [2, 3, 5, 6]. The GTA modelling requires the large and complex systems, such as the steam power plant network, to be reduced and divided into sub-systems for convenience of the analysis procedure. The GTA model simulates the inheritances and interdependencies of the sub-systems of the coal-fired generating station in addition to giving a quantitative measure of the system reliability. The GTA procedure is composed of three steps namely: (1) digraph representation; (2) matrix representation; and (3) development of a permanent structure function. The quantitative measure of the steam power plant system reliability enables the design engineer to determine the similarity or dissimilarity between the present reliability and the design value.

The GTA procedure is used here to model the entire system of a coal-fired generating station, as shown in **Figure 1**. The system is divided into six sub-systems ($N_i : i = 1, 2, \ldots.6$) which are given below [2, 3]:

N1: The coal system;

N2: The boiler system;

N3: The steam turbine;

N4: The boiler feed pump;

N5: The cooling system; and

N6: The generator.

The identified and above-mentioned sub-systems for the steam power plant of **Figure 1** are displayed in **Figure 2** [2, 3]. The discourse on the said six sub-systems follows in Section2.

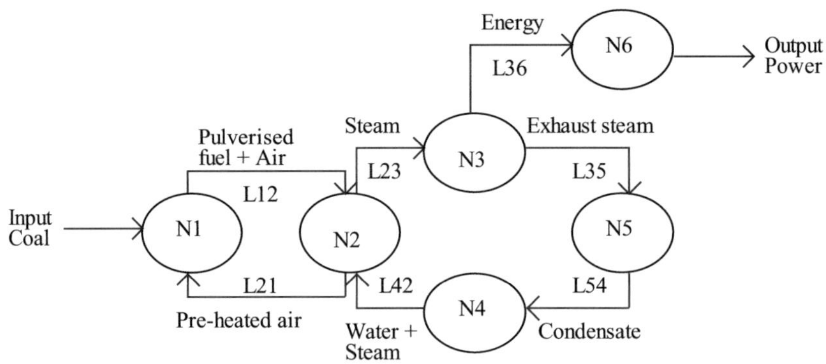

Figure 2. System structure digraph for a coal-fired generation station.

2. System structure function for the coal-fired generating station

In this research, consideration is made of systems of components that satisfy the following hypothesis ([2, 3], [4], pp. 123):

Systems that are composed of n components are denoted as systems of order n. The constituent components are assumed to be numbered consecutively from 1 to n. The study is confined to situations where it suffices to distinguish between only two states, a functioning state and a failed state. The preceding study limitation applies to each component as well as to the system itself. The state of component i, $i = 1, 2, \ldots. n$ can then be described by a binary[1] variable (function) x_i, where:

$$x_i = \begin{cases} 1, & \textit{if component i is functioning} \\ 0, & \textit{if component i is in a failed state} \end{cases} \tag{4}$$

Then, vector $X = (x_1, x_2, \ldots., x_n)$ denotes the states of all components. Vector $X = (x_1, x_2, \ldots., x_n)$ is known as the component *state vector*. In addition, the assumption is that if one knows all of the n components' states, then it follows that they also know the state of the system, that is whether it is functioning or failed. The system state is determined completely by and is an inevitable consequence of the states of the components that constitute the system.

In a similar way, the system state can then be delineated using a binary function: $\varnothing(X) = \varnothing(x_1, x_2, \ldots., x_n)$, where:

$$\varnothing(\mathrm{X}) = \begin{cases} 1, & \textit{if the system is functioning} \\ 0, & \textit{if the system is in a failed state} \end{cases} \tag{5}$$

and where $\varnothing(\mathrm{X})$ is called the *structure function* of the system or just the *structure* (e.g., the system structure digraph for a coal-fired generation station as shown in **Figure 2**). Each unique system corresponds to a unique structure function $\varnothing(\mathrm{X})$. Thus, one also talks about *structures* instead of *systems*.

The performance of a coal fired generating power station is a function of its basic *structure* (i.e., the layout and design), availability, safety and security, dependability, and other regulatory aspects. Availability here is defined as the ability of an item (under combined aspects of its reliability, maintainability, and maintenance support) to perform its required function at a stated instant of time or over a stated period of time (BS 4778) ([4], pp. 599). Understanding of its *structure function* will help in the improvement in performance, design, and maintenance planning. A mathematical model using the graph theory and matrix method is developed to evaluate the performance of a coal-based steam power plant in the subsequent sections.

[1]In this context a binary variable (function) is a variable (function) that can take only the two values, 0 or 1.

3. Graphical representation of the coal-fired generating station

In general, networks are any systems which admit abstract mathematical representations as graphs. The nodes (vertices) of these networks indicate these systems' components [2, 3]. The occurrence of a relation or intercommunication in the midst of the components in these networks is represented by the set of connecting links (edges). It can be noted that the preceding high level of abstraction can, in general, be applied to wide-ranging systems. Consequently, within that sense, a theoretical framework is provided by networks. This theoretical framework enables convenience of conceptual representation of system relations in which characterization at system level provides for component interactions mapping.

The thermal power plant system as shown in **Figure 1** is represented in the form of a graph $G = (N,\ L)$ of **Figure 2**, where N is the set of nodes (or vertices) and L the set of links (or edges) [2, 3]. Let each of the six sub-systems of the generating station be denoted by nodes $N_i's(i = 1,\ 2,\ \ldots.\ 6)$, and the interconnection between the systems (N_i, N_j) is represented by links $L_{ij}'s\ (i, j = 1,\ 2,\ \ldots.\ 6\ and\ i \neq j)$ joining nodes N_i and N_j. The flow of heat and energy, steam, water, pulverized fuel, and pre-heated air connects all the six sub-systems. Nodes and links aid in illustrating this flow in **Figure 2**. When the thermal power plant is graphically represented, this then is termed as the system *structure* function (i.e., as discussed in Section2) [2–4].

When the links (arcs/edges) can be traversed in both directions, the graph is undirected. On the other hand, the graph is directed if the links (arcs/edges) can be traversed only in one direction indicated by an arrow. If an undirected graph has no self-loops the presence of at least one link per node guarantees that all the nodes are connected [2, 3]. Practical structures are in general substantially more connected as compared to this minimal threshold. Consequently, there exist numerous paths between any node pair. For directed graphs, the connectivity property is unwieldy because the nodes can relate to any of the following three of categories: (1) the nodes that are strongly connected (i.e., for this subset the nodes can be arrived at from any other node that is a member of the subset. The access is through following the direction of the links); (2) the transient nodes that only have outgoing links. Therefore, transient nodes cannot be accessed from any other node; and (3) the absorbing nodes that only have ingoing links. Thus, once reached, the absorbing nodes cannot be left [2, 3].

The main intuitive and illustrative tool to be used, for the coal-fired generating station system structure function, is the directed graph [2, 3]. Intuitively the directed graph can be considered as a set of points (or vertices/nodes) with arrows (or arcs/links/edges) joining some of the points [7]. An arc may be labelled. Conventionally, a digraph consists of a set of vertices (nodes/points), V(N) and a subset of ordered pairs of arrows called the arcs (links). The labelling of the digraph is a function from the arcs (links) to the real numbers. One can visualize a labelled digraph by considering the vertices (nodes) as points with arcs (links) as arrows going from vertex (node) i to vertex (node) j whenever (i, j) belongs to the sets of arcs (links). The ith vertex (node) of the arc (link) (i, j) is called its initial vertex (node), while the jth vertex (node) is called its terminal vertex (node). The arc (link) is then given a label which is the representation of that arc (link) under the labelling function. When the initial and terminal vertices (nodes) are identical, the arc (link) is called a loop. It can be noted that sometimes an arc (link) originates from its initial vertex (node) and that it terminates into its terminal vertex (node). The number of arcs (links) that originate from a vertex (node) is called its out degree, and the number of arcs (links) that terminates into that vertex is called its in degree.

4. Reliability assessment of the coal-fired generating station

According to Ref. [8] as cited in ([4], pp. 5): "When the words are used sloppily, concepts become fuzzy, thinking is muddled, communication is ambiguous, and decisions and actions are suboptimal, to say the least." Therefore, it follows from this saying that a precise definition of reliability and some associated concepts is needed. There are several definitions of what reliability is. In this study reliability is defined as the ability of an item to perform a required function, under given environmental and operational conditions and for a stated period of time (ISO 8402) ([4], pp. 5). The terms in this definition are explained as follows [4]:

- The word "item" within the context of this research stands for any component, sub-system, or system which can be envisaged as an entity;
- A requisite utility could be a single utility or a combination of utilities which are needed to carry out a service;
- Components, sub-systems, and systems are herein referred to as technical items. These items have been designed so as to carry out one or more (needed) functions. Furthermore, some of these functions are active, whereas others are passive. An example of a passive function is containment of fluid in a pipeline. Involved and intricate systems (e.g., a steam power plant) usually have a wide range of required functions. To assess the reliability (e.g., of a steam power plant), one needs to state all the function(s) required for consideration; and
- A hardware item has to be reliable. This reliability level is satisfied when the item in question performs above the initial factory specification. Furthermore, it is imperative that the hardware item operates satisfactorily in the actual environment for which it is designed, for a specified period of time.

Reliability has two abstract meanings; probabilistic and deterministic. The probabilistic approach is based upon statistical failure modelling, without researching and itemizing causes of failure. On the other hand, the deterministic approach focuses on understanding how and why a component or system has failed, and how it can be designed, repaired and tested to prevent such failure from occurrence or recurrence. In the present analysis, the probabilistic approach in conjunction with the graph theory is applied for the steam power plant.

Let $R_i(i = 1,\ 2,\ \ldots.\ 6)$ denotes the reliability of node N_i and $r_{ij}(i, j = 1,\ 2,\ \ldots.\ 6\ and\ i \neq j)$, the reliability of the link (or interconnection) between the nodes, N_i and N_j [2, 3]. Consequently, associating reliability to the system structure of **Figure 2** results in the system reliability graph modelling. The system reliability graph for the coal-fired generating station corresponding to its abridged system structure graph is obtained by associating R_i with N_i and r_{ij} with $L_{ij},$ and this is shown in **Figure 3** [2, 3].

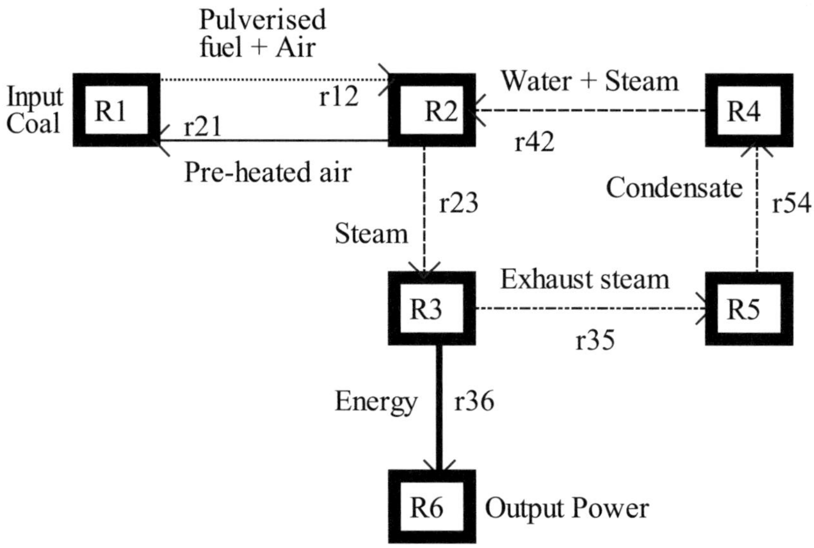

Figure 3. System reliability digraph for a coal-fired generating station.

The reliability structure function of the system of **Figure 3** is estimated by computing the graph's characteristic polynomial [2, 3]. Each and every finite directed graph has a characteristic polynomial [9]. In the beginning, the characteristic polynomial was believed to be a complete invariant, or unique to a graph and all its isomorphism. Later, it was discovered that there are cases where structurally different graphs share the same characteristic polynomial [9]. Although there are cases where structurally different graphs share the same characteristic polynomial, characteristic polynomials are highly studied because they provide much information about a graph in concise format. Characteristic polynomials are useful in steam power plants, mathematics, chemistry, economics, and physics, among others. For example, a graph's spectrum (i.e., the roots of its characteristic polynomial) has significance in the atomic structure.

There exist several ways of determining the characteristic polynomial of a graph with n vertices. In this research, we utilize three methods, namely: (1) the linearly independent cycles; (2) the formula called the figure equation; and (3) the adjacent matrix method. These three methods are in turn employed to estimate the system reliability structure function. Estimating reliability is always an imperfect endeavour (i.e., the reliability (estimate) ranges from the lower bound to the upper bound) and hence the use of these three methods for comparison purposes [2, 3]. The three methods are discussed in the subsequent sections.

4.1. The linearly independent cycles method

In the linearly independent cycles procedure, the structure function is characterized by the presence of a sufficient number of certain cycles which have the property that they are linearly independent [2, 3, 10]. These cycles are denoted in a matrix form (i.e., the cycles are denoted by the links (edges/arcs) present in them). Let $A = (a_{ij})$ of dimension $L \times L$, where L denotes the number of links and an element a_{ij} denotes whether link j is present in cycle i or not. When the cycle accumulates the link metrics on its path, then the value observed at m is the sum of the link metrics. Furthermore, the assumption here is that the monitor node is responsible for starting and terminating probes. In the preceding case, m is the monitor node.

One assigns B to denote $(L \times 1)$ a column matrix which contains the accumulated metrics that correspond to the linearly independent cycles. Furthermore, one assigns x to denote $(L \times 1)$ a column matrix that contains the link variables which one has to identify. One's ultimate goal is to solve $Ax = B$ which represents a system of linear equations. For one to be able to uniquely determine x, A has to be invertible. A is also referred to as identifiable, because it has full rank. Cycles that make up such a matrix (i.e., matrix A) are referred to as linearly independent cycles. All link metrics can be uniquely identified by solving Eq. (6) [2, 3, 10]:

$$x = A^{-1}B \tag{6}$$

4.2. The figure equation method

The figure equation procedure provides a direct link between a graph's structure function and the coefficients of its characteristic polynomial. Unlike the linearly independent cycles and the adjacent matrix methodologies, the figure equation method does not use determinants but calculates the characteristic polynomial of any graph by counting the cycles in the graph [2, 3, 9]. In this procedure, coefficients of the graphs' characteristic polynomials are calculated. The calculations are done when one considers the set of linearly-directed sub-graphs of a corresponding length. i nodes and i links constitute a linearly-directed sub-graph of length i in such a way that each node bears in degree and out degree of one (1).

The figure equation states that for any graph $G = (N, L)$ with n vertices, the characteristic polynomial is [2, 3, 9] as follows:

$$X(G) = x^n + c_1x^{n-1} + \ldots + c_{n-1}x + c_n \tag{7}$$

such that for $1 \le i \le n$, the coefficient is as follows:

$$c_i = \sum_{L \in L_i} (-1)^{P(L)} \tag{8}$$

where L_i is the set of all linearly directed subgraphs of G and $P(L)$ is the number of *linearly directed cycles*, or the number of pieces in L.

4.3. The adjacent matrix method

An adjacency matrix tells which vertex (node) in the graph is connected to which. The adjacency matrix of the system reliability digraph for a coal-fired generating station is defined to represent the steam power plant. The adjacent matrix should be defined such that it incorporates the structural information of the components and sub-systems (i.e., of the steam power plant) and interconnections between them [2–4].

The system structure matrix of the steam power plant is defined as follows [2, 3]. Here, a general case of a macro level coal-fired generating station with N sub-systems is considered. Thus, leading to a symmetric adjacency, matrix $\{0, 1\}$ of order $N \times N$. $c_{ij}(i, j = 1,\ 2,\ \ldots.6\ and\ i \neq j)$ represents the connectivity between node (vertex) i and j such that $c_{ij} = 1$ if node (vertex) i is connected to node (vertex) j. In the system reliability digraph of **Figure 3**, this is represented by the link (edge) reliability r_{ij} between nodes i and j. c_{ij} is equal to zero otherwise. Thus, $c_{ij} = 0$ for all i, as a node (vertex) cannot be connected to itself. In the case, where the node (vertex) is connected to itself, $c_{ij} = 1$. This implies a self-loop at node (vertex) i in the graph.

Each row or column of the system structure matrix corresponds to a node (vertex). The six sub-systems of the system reliability digraph of **Figure 3** correspond to the six columns or rows of this matrix [2, 3]. The off-diagonal matrix elements, c_{ij}, represent a connection between nodes (vertices) i and j. In the adjacent matrix $c_{ij} \neq c_{ji}$, as only directional connections between nodes (vertices) are considered. The characteristic polynomial of the graph is the characteristic polynomial of its adjacency matrix. The determinant of the characteristic system reliability matrix is called the characteristic system reliability polynomial.

The reliability of the system is estimated by obtaining the determinant of the characteristic system reliability matrix as follows [2–6]:

$$R_{system} = det\{RI - A_{adjacent}\} \tag{9}$$

where R represents the reliability of the nodes constituting the system; I is the node identity matrix; and $A_{adjacent}$ is the system structure adjacency matrix.

5. Illustration of the system reliability methodology assessment

5.1. Illustration of the linearly independent cycles procedure

The linearly independent cycles (LIC) method computes the list of cycles (top) and the corresponding cycle-link matrix (bottom) as shown in **Figure 4**. The two (2) cycles computed in **Figure 4** are linearly independent. Thus, all link metrics may be identified. We use node R_1 (see **Figure 3**) as the monitor that can start and terminate probes. The reliability of the structure of **Figure 3** is determined using Eq. (6) as [2–6]:

$$R_{system} = \det\{(R \times eye(7)) - A_{LIC}\} = R^7 - 2R^6 \tag{10}$$

For constant unit failure rate, substituting $R(t) = e^{-\lambda t}$ into Eq. (10) yields:

$$R_{system}(t) = e^{-7\lambda t} - 2e^{-6\lambda t} \tag{11}$$

where $R_{system}(t)$ is the coal-fired generating station reliability at time t and λ is the unit constant failure rate.

$$r_{12} \rightarrow r_{21}$$

$$r_{12} \rightarrow r_{23} \rightarrow r_{35} \rightarrow r_{54} \rightarrow r_{42} \rightarrow r_{21}$$

$$A_{LIC} = \begin{pmatrix} r_{12} r_{21} r_{23} r_{35} r_{36} r_{42} r_{54} \\ 1\ 1\ 0\ 0\ 0\ 0\ 0 \\ 1\ 1\ 1\ 1\ 0\ 1\ 1 \\ 0\ 0\ 0\ 0\ 0\ 0\ 0 \\ 0\ 0\ 0\ 0\ 0\ 0\ 0 \\ 0\ 0\ 0\ 0\ 0\ 0\ 0 \\ 0\ 0\ 0\ 0\ 0\ 0\ 0 \\ 0\ 0\ 0\ 0\ 0\ 0\ 0 \end{pmatrix}$$

Figure 4. List of linearly independent cycles (top) and the corresponding cycle-link matrix (bottom).

5.2. Illustration of the figure equation procedure

The calculation of characteristic polynomials using the figure procedure is relatively unfamiliar [9]. In the figure equation procedure, coefficients of the graphs' characteristic polynomials are calculated. The calculations are done when one considers the set of linearly-directed subgraphs of a corresponding length. i nodes and i links constitute a linearly-directed sub-graph of length i in such a way that each node bears in degree and out degree of one (1) [9].

From Section4.2, we recall that a factor F of a digraph H is a subgraph containing all the vertices of H in which each vertex (node) has both in degree and out degree equal to one. In other words, it consists of a collection of disjoint cycles that go through each vertex (node) of H. The number of cycles in the factor F is denoted $n(F)$. If the digraph H is labelled, then $W(F)$ denotes the weight of the factor [7]. Given the digraph H of order n; F its factor; F its set of all linearly directed subgraphs; $n(F)$ the number of cycles in the factor F; $W(F)$ the weight of the factor F, then the coefficients, c_i of H are determined as follows:

$$c_i = \sum_{F \in F} (-1)^{n(F)} W(F) \tag{12}$$

where $1 \le i \le n$ and in this study we assume $W(F) = 1$.

Using the figure equation formulae (Eqs. (7), (8), and (12)), we get the list of sub-graphs (left) and the corresponding coefficients (right) as shown in **Table 1**. From the information as shown in **Table 1**, the characteristic polynomial of the structure in **Figure 3** is as follows:

Cycles in the sub-graph	Coefficients
$r_{12} \rightarrow r_{21}$	2 *pieces*, $c_1 = (-1)^2 = 1$
$r_{12} \rightarrow r_{23} \rightarrow r_{35} \rightarrow r_{54} \rightarrow r_{42} \rightarrow r_{21}$	6 *pieces*, $c_2 = (-1)^6 = 1$
$r_{23} \rightarrow r_{35} \rightarrow r_{54} \rightarrow r_{42}$	4 *pieces*, $c_3 = (-1)^4 = 1$

Table 1. Linearly directed cycles and their coefficients.

$$R(G) = x^6 + x^5 + x^4 + x^3 \tag{13}$$

For constant unit failure rate, substituting $R(t) = e^{-\lambda t}$ into Eq. (13) yields:

$$R_{system}(t) = e^{-6\lambda t} + e^{-5\lambda t} + e^{-4\lambda t} + e^{-3\lambda t} \tag{14}$$

where $R_{system}(t)$ is the coal-fired generating station reliability at time t and λ is the unit constant failure rate.

5.3. Illustration of the adjacent matrix procedure

The digraph for the coal-fired generating station of **Figure 3** characterizes the visual representation of the system and its interdependence [2, 3]. The adjacent matrix procedure converts the digraph into a mathematical form, and the structure function is a mathematical model that helps to determine the reliability index [2, 3]. It may be noted here that the development of a structure function is not merely the determinant of the matrix. The structure function is developed in such a manner that no information regarding the system reliability is lost [2, 3]. For this purpose, a step-by-step procedure is proposed in Section 4.3.

Using the adjacent matrix method, the reliability of the structure of **Figure 3** is determined using Eq. (9) as [2–6]:

$$R_{system} = \det\left\{\left(\begin{bmatrix} R_1 & 0 & 0 & 0 & 0 & 0 \\ 0 & R_2 & 0 & 0 & 0 & 0 \\ 0 & 0 & R_3 & 0 & 0 & 0 \\ 0 & 0 & 0 & R_4 & 0 & 0 \\ 0 & 0 & 0 & 0 & R_5 & 0 \\ 0 & 0 & 0 & 0 & 0 & R_6 \end{bmatrix} \times \begin{bmatrix} 1 & 0 & 0 & 0 & 0 & 0 \\ 0 & 1 & 0 & 0 & 0 & 0 \\ 0 & 0 & 1 & 0 & 0 & 0 \\ 0 & 0 & 0 & 1 & 0 & 0 \\ 0 & 0 & 0 & 0 & 1 & 0 \\ 0 & 0 & 0 & 0 & 0 & 1 \end{bmatrix}\right) - \left(\begin{bmatrix} 0 & r_{12} & 0 & 0 & 0 & 0 \\ r_{21} & 0 & r_{23} & 0 & 0 & 0 \\ 0 & 0 & 0 & 0 & r_{35} & r_{36} \\ 0 & r_{42} & 0 & 0 & 0 & 0 \\ 0 & 0 & 0 & r_{54} & 0 & 0 \\ 0 & 0 & 0 & 0 & 0 & 0 \end{bmatrix}\right)\right\} \tag{15}$$

$$= R_6\,(R_3\,R_4\,R_5\,r_{12}\,r_{21} + R_1\,r_{23}\,r_{35}\,r_{42}\,r_{54} - R_1\,R_2\,R_3\,R_4\,R_5)$$

For identical units (i.e., for illustration purposes only) (i.e., $R_1 = R_2 = R_3 = R_4 = R_5 = R_6 = R$) with the link reliability assumed to be unity (i.e., $r_{ij} = 1$), Eq. (15) simplifies to:

$$R_{system} = \left|\left(R \times (-R^5 + R^3 + R)\right)\right| \tag{16}$$

where $|\bullet|$ denotes the absolute value and det is the matrix determinant.

http://dx.doi.org/10.5772/intechopen.68285

For constant unit failure rate, substituting $R(t) = e^{-\lambda t}$ into Eq. (16) yields:

$$R_{system}(t) = |e^{-\lambda t}(e^{-\lambda t} + e^{-3\lambda t} - e^{-5\lambda t})| \quad (17)$$

where $R_{system}(t)$ is the coal-fired generating station reliability at time t and λ is the unit constant failure rate.

6. Results

6.1. The approximate system reliability

The system reliability value of the coal-fired generating station of **Figure 1** as illustrated by Eqs. (11), (14), and (17), respectively, is plotted as shown in **Figure 5**. It should be noted that the results in **Figure 5** have been obtained assuming that the components are identical and assuming constant unit failure rate. This is only for the example to illustrate the methods. In practice, components are not identical, and different failure rates could be used for each component, which is the real-life scenario. The results of the approximation of the system reliability for the three methods which are shown in **Figure 5** have three bounds: (1) the lower; (2) the in-between; (3) and the upper. The lower bound is given by the linearly independent cycles method; the in-between bound by the adjacent matrix method and then the upper bound by the figure equation method. **Figure 5** reveals that the approximate system reliability value starts to decrease gradually with time as is expected. There are various reasons for this

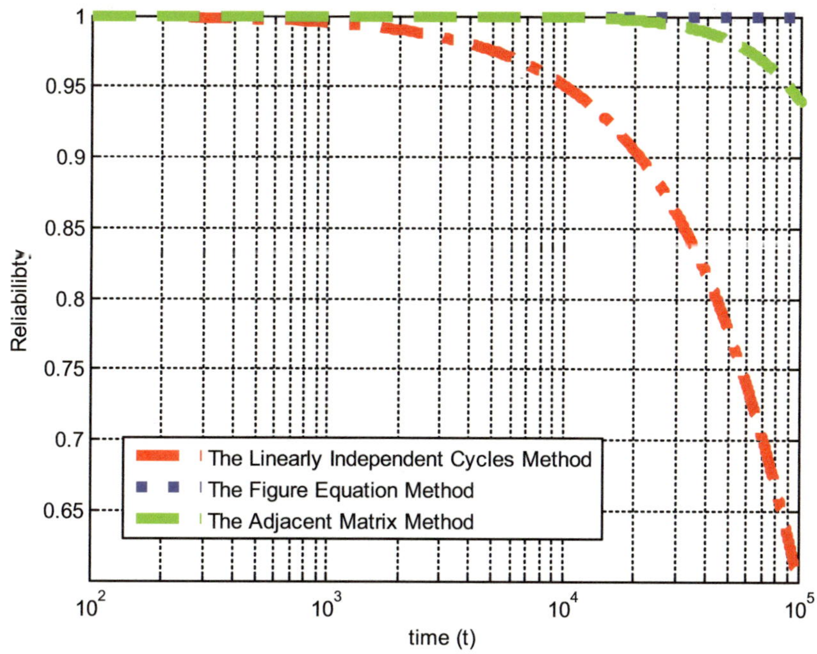

Figure 5. Coal-fired generating station system reliability.

phenomenon. Some of the possible causes are: (1) the ageing effects of the system; and (2) the unavailability of some of the components that constitute the sub-systems of the thermal power plant. The thermal power plant actual or real-time performance system reliability, when available, is compared with the design system reliability (approximate/estimate) by way of the graph-theoretical analysis. The main objective of this comparison is to optimize the design system reliability (estimate).

6.2. The Birnbaum's measure of structural importance

In 1969, Birnbaum advocated the measure for the structural importance of component i as follows [3, 4]:

$$B_{\varnothing}(i) = \frac{\eta_{\varnothing}(i)}{2^{n-1}} \tag{18}$$

The measure of structural importance, $B_{\varnothing}(i)$, proposed by Birnbaum and illustrated in Eq. (18), shows the comparative portion of the 2^{n-1} possible state vectors, that is, $(\cdot_i, x)$.These possible sate vectors form critical path vectors for component i. Birnbaum's measure of structural importance, as expressed in Eq. (18), can be used to partially rank components constituting a system in accordance with the size of $B_{\varnothing}(i)$.

Figure 6 shows the results for Birnbaum's structural importance for the components in the coal-fired generating station of **Figure 1**. **Figure 6** illustrates how much worse the coal-fired generating station system reliability would be if component i would fail. **Figure 6** shows that eight components (N_2, N_6, L_{12}, L_{21}, L_{23}, L_{35}, L_{42}, and L_{54}) have the same Birnbaum structural importance (i.e., 3.12%) with four components (N_1, N_3, N_4, and N_5) having a negligibly small Birnbaum structural importance (i.e., 0%). Birnbaum's measure of structural importance only takes into account the system structure function and not the lifetime distributions of the components. Therefore, it is relatively easy to calculate and is in general used in the design

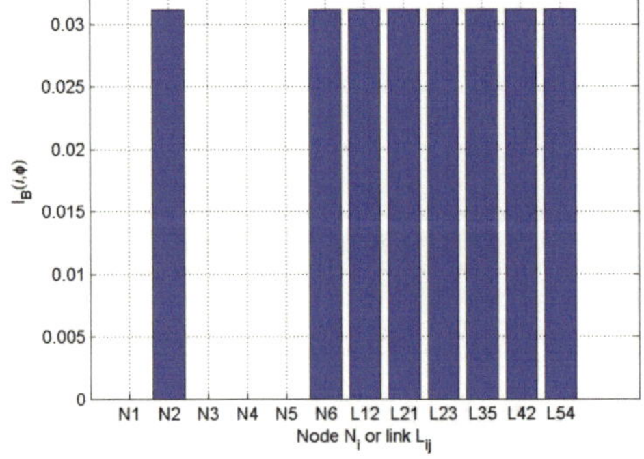

Figure 6. Birnbaum's measure of structural importance for components in the thermal power plant.

phase or when the lifetime distributions of components are not known. Birnbaum's measure of structural importance is also an alternative when the more advanced measures would be too time-consuming to compute or difficult to use.

6.3. The Birnbaum's measure of reliability importance

In 1969, Birnbaum proposed the measure of the reliability importance of component i at time t as follows [3, 4]:

$$I^B(i|t) = \frac{\partial h\left(p(t)\right)}{\partial p_i(t)} \; for\; i = 1,\; 2,\; \ldots, n \tag{19}$$

In order to obtain Birnbaum's measure of the reliability importance of component i at time t. A partial derivative of the system reliability with respect to $p_i(t)$ is taken. Birnbaum's measure of the reliability importance is a specific case of sensitivity analysis that was used in various engineering applications for ages [4]. A large $I^B(i|t)$ results in a comparatively large change in the system reliability at time t, for a small change in the reliability of component i. Birnbaum's measure of the reliability importance depends on the component reliabilities at various points in time. Therefore, it gives perhaps a more global view of component importance.

Figures 7 and **8** show that all components except (N_1, N_3, N_4, and N_5) have the same Birnbaum's measure of the reliability importance. We deduce how the components behave with respect to time in **Figures 7** and **8**.

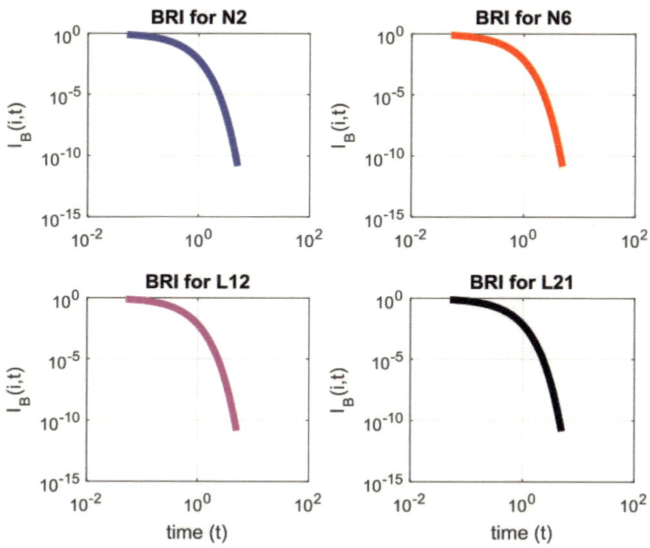

Figure 7. Birnbaum's measure of reliability importance for components: N_2, N_6, L_{12}, and L_{21}.

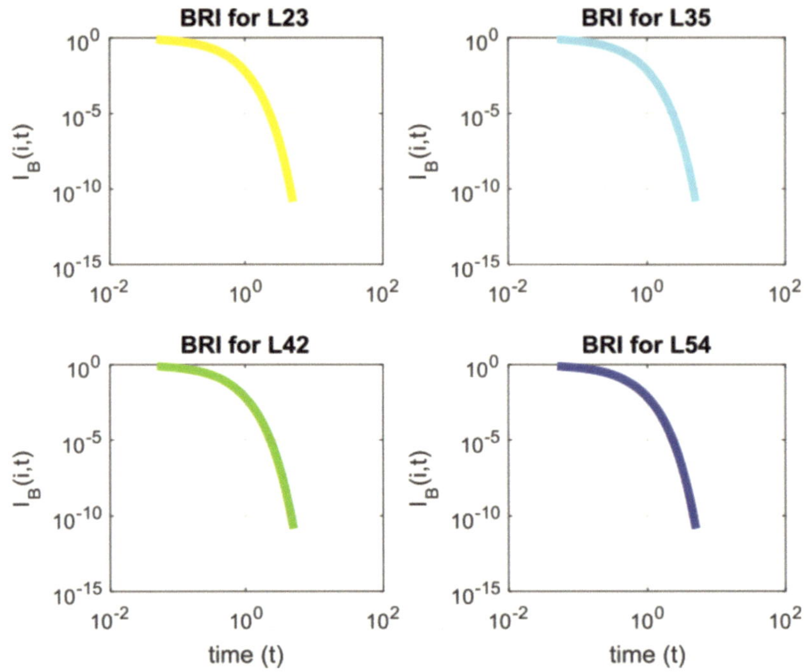

Figure 8. Birnbaum's measure of reliability importance for components: L_{23}, L_{35}, L_{42}, and L_{54}.

7. Conclusion

The graph-theoretical analysis procedure was used for the analysis of several reliability parameters of a steam power plant, in this chapter. **Figure 5** illustrates the system reliability as a function of time with failure rates assumed to follow exponential time distributions. It is worth noting that the results of **Figure 5** have been attained assuming identical components and constant unit failure rates. This is only for the example to illustrate the methods. In practice, components are not identical and different failure rates could be used for each component, which is the real-life scenario.

The structure function model for the steam power plant developed in this chapter represents its structural information, including its systems, their sub-systems, their components and their interconnections. The procedure transforms a real-life steam power plant into the following representations: its block (see **Figure 1**); its system structure digraph (see **Figure 2**); and then finally its system reliability digraph (see **Figure 3**). The structure function of the coal-fired generating station embodies all probable composites of its components and sub-systems at a specific state of hierarchy. These composites and interactions create a method which can be used to analyse the structure and the function of the parameters that are dependent on the structure. The said methodology allows for either a top-down or bottom-up analysis and design of various systems, sub-systems, and their interconnections. The model enables one to determine optimal maintenance strategies that will ascertain upper bound reliability of the thermal power station. Power plant managers can benefit from use of the model for the analysis of the reliability of the thermal power plants. The procedure permits changes in the model to make it plant-specific as well as design-specific.

In general, a collection of components performing a specific task or function is referred to as a system. It goes without saying that in a system, some components are more important for the system reliability than others. For example, in a system, if a component is in series with the rest of the components, it is a cut set of order one (1). A component which is a cut set of order one (1) is in general more important than a component that is a member of a cut set of higher order ([4]: pp. 149–150). In Section 6, Birnbaum's measure of structural importance and Birnbaum's measure of the reliability importance have been defined and discussed. Component importance measures may be used to rank the components, that is, to arrange the components in ascending or descending order of importance. Component importance measures may also be used for classification of importance, that is, to allocate the components into two or more groups, according to some pre-set criteria.

Systems usually consist of multiple components. The components constituting a system are not necessarily equally important for the performance (reliability, availability, risk, and throughput) of the constituent system. Ordinarily limited resources are available to design, enhance and/or maintain such a system efficiently. Nevertheless, for complex and large systems, it may be too tedious, or not even possible, to develop a formal optimal strategy. In analogous situations, it is advantageous for one to allocate resources in accordance of how important the components are to the system. Furthermore, it is desirable to concentrate the resources on the subset of components that are most important to the system ([11]: pp. 49–53). Therefore, the notion of component importance measures (also called sensitivity).

A basic problem that faces the reliability engineer in attempting to achieve maximum reliability for a large and complex system is that of evaluating the relative importance of the various components constituting the system. Thus, in reliability, a component importance measure evaluates the relative importance of individual components or group of components in that system. This relative importance can be determined based on the system structure, component reliability and/or component lifetime distributions. Measuring the relative importance of components may allow the engineer to: (1) determine which of these components deserve additional research and warrant development in order to improve the overall system reliability under cost (and/or effort) constraints; and (2) find the component that caused the failure of a system. By using importance measures, it is possible to draw conclusions about which components are the most important to improve in order to achieve better reliability of the whole system. In Sections 6.2 and 6.3, we have provided Birnbaum's measure of structural importance and Birnbaum's measure of the reliability importance for that specific purpose and given the results thereof.

Author details

Bernard Tonderayi Mangara

Address all correspondence to: aragnam@gmail.com

Central University of Technology, Free State (CUT), Bloemfontein, South Africa

References

[1] Weedy BM, Cory BJ, Jenkins N, Ekanayake JB, Strbac G. Electric Power Systems. 5th ed. Chichester, United Kingdom: John Wiley & Sons Ltd; 2012. p. 496.

[2] Mangara BT, Finkelstein M. On the improvement of steam power plant system reliability. In: 40th Annual Conference of the IEEE Industrial Electronics Society (IECON 2014); 29 October–01 November; Sheraton Hotel, Dallas, Texas, USA. New York: IEEE; 2014. pp. 5389–5393. DOI: 978-1-4799-4033-2/14

[3] Mangara BT. On some methods of reliability improvement of engineering systems [thesis]. Bloemfontein, Free State, South Africa: University of the Free State; 2015. p. 155.

[4] Rausand M, Hoyland A. System Reliability Theory: Models, Statistical Methods, and Applications. 2nd ed. New Jersey, USA: John Wiley & Sons, Inc.; 2004. p. 636.

[5] Mohan M, Gandhi OP, Agrawal VP. Real-time reliability index of a steam power plant. Proceedings of the Institution of Mechanical Engineers, Part A: Journal of Power and Energy. 2008;**222**:355–369.

[6] Dev, N, Samsher, Kachhwaha SS, Attri R. Development of reliability index for combined cycle power plant using graph theoretic approach. Ain Shames Engineering Journal. 2014; **5**(1):193–203.

[7] Doob M. Applications of graph theory in linear algebra. Mathematics Magazine. 1984;**57** (2):67–76.

[8] Kaplan S. Bayes is for eagles. IEEE Transactions on Reliability. 1990;**39**:130–131.

[9] Coon T. Combinatorics of the figure equation on directed graphs. Rose-Hulman Institute of Technology: Undergraduate Math Journal. 2006;**7**(2); https://www.rose-hulman.edu/mathjournal/archives/2006/vol7-n2/paper17/v7n2-17pd.pdf: 1–13.

[10] Gopalan A, Ramasubramanian S. On identifying additive link metrics using linearly independent cycles and paths. IEEE/ACM Transactions on Networking. 2012;**20**(3):906–916.

[11] Kuo W, Zhu X. Importance Measures in Reliability, Risk, and Optimisation: Principles and Applications. Chichester, UK: John Wiley & Sons, Ltd; 2012. p. 440.

Section 3

Nuclear Power Plant Safety and Technology Improvement

Chapter 5

Risk Assessment of NPP Safety in Case of Emergency Situations on Technology

Juraj Králik

Additional information is available at the end of the chapter

http://dx.doi.org/10.5772/intechopen.68772

Abstract

The last accidents of the nuclear power plant (NPP) in Chernobyl and Fukushima give us the new inspiration to verify the safety level of the NPP structures. This paper presents the new requirements to test the safety and reliability of the NPP structures due to the recent accidents in the world. The IAEA in Vienna required in the document 'Stress tests' the verification of the safety of the NPP structures under impact of the extreme loads as the earthquakes, the extreme climatic actions and the technology accidents. The new recommendations to load combinations and design criteria were defined. The risk assessment to verify the safety and reliability of the NPP structures based on probabilistic and nonlinear analysis is presented. The uncertainties of material model (behaviour of the reinforcement and liner, concrete cracking and crushing), degradation effects, the loads level (dead and live loads, extreme climatic and accidental temperature and overpressure) as well as other effects following from the inaccuracy of the calculated model and numerical methods were taken into account in the response surface method (RSM) method. The results of the deterministic and probabilistic analysis of the NPP structures are presented.

Keywords: risk, stress test, accident, safety, probability, nonlinearity, VVER, NPP

1. Introduction

The nuclear technology gives us the perspective of the effective use of natural resources of energy, but on the other hand, it is a risk for the environment [1–4]. The first accident in nuclear research facilities dates back to 21 May 1946. A nuclear critical accident occurred at the Los Alamos Scientific Laboratory in New Mexico. Severe problems arrived after an accident in Chernobyl Nuclear Power Plant in Ukraine on 26 April 1986. The last massive nuclear power plant (NPP) accident was on 11 March 2011 in Fukushima [5–7]. Following a major earthquake (a magnitude 7.1), a 15-m tsunami disabled the power supply and cooling of three Fukushima Daiichi reactors, hence causing a nuclear accident.

In view of the analysis results of the Fukushima accident [5, 6], the owner and operator of nuclear power plants (NPPs) in Slovakia re-evaluated the safety and reliability of structures and technology of all objects, according to the recommendations in "Stress tests" on all units in operation or under construction [7]. The main plant technological equipment (except for main circulation pumps) for all units was manufactured either in Czech Republic or in Slovakia [3].

In the process of design, construction and operation of plants, significant improvements in safety were made compared to the original project, including enhanced resistance to external hazards [1–3, 8]. Lately, in addition to previous safety improvements, many proposals have been implemented in the reconstruction process of NPP structures with reactor VVER 440/V213 in Slovakia for mitigation of severe accidents [3]. Fifty-one per cent of the overall production of electricity in SR comes from the nuclear power plants (**Table 1**).

The NPP buildings with the reactor VVER 440/213 consist of the turbine hall, middle building, reactor building and bubble condenser [1]. The NPP containment is limited by the area of reactor building and bubble tower (**Figure 1**). In the case of the loss-of-coolant accident (LOCA), the

Source	2007	%	2008	%	2009	%
Nuclear power plants	15,335	51.80	16,704	56.00	14,081	51.42
Thermal power plants	5421	18.30	5647	18.93	4768	17.41
Hydroelectric power plants	4485	15.10	4284	14.36	4662	17.02
Other power plants	2666	9.00	2674	8.96	2563	9.36
Import saldo	1725	5.80	521	1.75	1312	4.79
Consumption SR	29,632	100.00	29,830	100.00	27,386	100.00
Production SR	27,907		29,309		26,074	

Table 1. Source ratio on annual electricity consumption in SR.

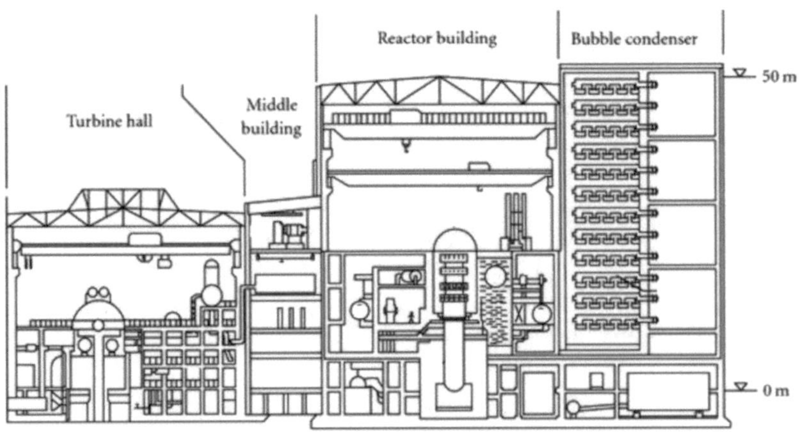

Figure 1. Section plane of the containment with reactor VVER440/213.

steam expands in the containment space, and the overpressure and temperature loads affect the containment structure [1]. The structural integrity and the tightness of the containment must be guaranteed in the case of the technology accident.

2. Safety assessment of the NPP risk

The International Atomic Energy Agency (IAEA) set up a programme [8] to give guidance to its member states on the many aspects of the safety of nuclear power reactors. The IAEA standards [8–10] and the US Nuclear Regulatory Commission (NRC) [4, 11] define the principal steps for the calculation of the risk of the NPP performance by simulations using Latin hypercube sampling (LHS) probabilistic method as follows:

1. Accident frequency (systems) analysis
2. Accident progression analysis
3. Radioactive material transport (source-term) analysis
4. Offsite consequence analysis
5. Risk integration.

The final stage of the probabilistic safety analysis (PSA) is the compilation of the outputs of the first four steps into an expression of risk. The risk integration is shown in matrix formulation in **Figure 2**. The approximate numbers of plant damage states (PDSs), accident progression bins (APBs), and source-term groups (STGs), and the number of consequences are presented in the different reports [1–3, 12] and standards [4, 10–13].

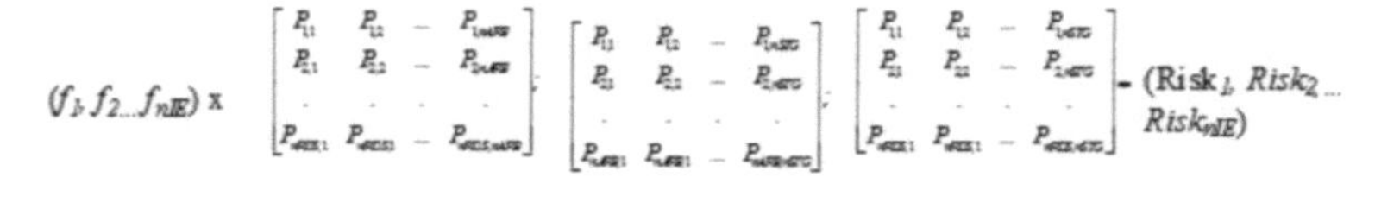

Figure 2. Scheme of Latin hypercube sampling.

3. Safety of the NPP structure resistance

In the case of the loss-of-coolant accident (LOCA) [3, 17, 18], the steam pressure expands from the reactor hall to the bubble condenser. Hence, the reactor hall and the bubble condenser are the critical structures of the NPP hermetic zone.

In the past, the calculation of the structural reliability for containment of the type of VVER 440/213 was carried out determining the probability density function for the ultimate pressure. A basis for the calculations consisted in results of linear and non-linear analysis, depending on whether the modelling considered the update of the properties.

The present work analyses the impact of combination of pressure load with thermal load that can arise in extreme situations related to severe accident progression. To achieve proper results, a detailed finite-element analysis of the concrete structure was carried out using ANSYS software and the programs CRACK [3, 14–18] were employed to solve this task. The basis for the probabilistic evaluation is reviewed, considering the uncertainties connected with loads and material properties. In the "Result" section, a comparison with linear evaluation is also mentioned.

3.1. Scenario of the hard accident

Most of the countries rate the safety of the NPP hermetic structures through the double-ended break guillotine test of the largest pipe in the reactor coolant system. During the design process of NPP structures with reactors VVER-440/230 type, the guillotine break of 2 × 500-mm pipes was considered as a beyond-design-basis accident (BDBA). The scenario of the hard accident is based on the assumption of the extreme situation where a loss of primary coolant accident (LOCA) is combined with the total loss of containment cooling system. During this situation, the hermetic zone cannot be available due to the external spraying of borated water. In practice, the contribution to heating from hydrogen explosion shall also be accounted.

In addition, the extremely climatic temperature (negative or positive) impacts the external slabs and walls of the hermetic zone. The temperature boundary conditions are defined to comply with the new revision of reference temperature provided by the Slovak hydro-meteorological institute (SHMU) [3] and relevant Eurocodes [22] for a return period of 10^4 year. Three types of the scenarios were considered (**Table 2**).

3.2. Steel and concrete material properties under high-temperature impact

The recapitulation of the research works and the standard recommendations for the steel and concrete under high-temperature effect are summarized in US NRC report [23] and Eurocodes [22].

Type	Duration	Over-pressure [kPa]	Extreme temperatures [°C]		
			Interior	Exterior	
				Max.	Min.
I.	1 day	150	127	42	−28
II.	7 days	250	150	42	−28
III.	1 year	–	80–120	42	−28

Table 2. The scenarios of the accidents in the hermetic zone.

The recommendations for the design of the structures are described in the US standards ACI [24], CEB-FIP Model Code [25] and Eurocodes [22]. The bonding phases of concrete are from the instable substance, which can be destructed at high temperature and their microstructures are changed.

The thermal conductivity λ_c of normal weight concrete may be determined between the lower and upper limits given hereafter. The upper limit has been derived from tests of steel-concrete composite structural elements. The CEB-FIP Model Code [25] and Eurocodes [22] define the stress-strain relationship for concrete and steel materials dependent on temperature θ for heating rates between 2 and 50 K/min. In the case of the concrete, the stress–strain diagram is divided into two regions. The concrete strength $\sigma_{c,\theta}$ increases in the first region and decreases in the second region (**Figure 3**).

The stress-strain relations $\sigma_{c,\theta} \approx \varepsilon_{c,\theta}$ in region I are defined in the following form:

$$\sigma_{c,\theta} = f_{c,\theta}\left[3\left(\frac{\varepsilon_{c,\theta}}{\varepsilon_{cu,\theta}}\right) \Big/ \left\{2+\left(\frac{\varepsilon_{c,\theta}}{\varepsilon_{cu,\theta}}\right)^3\right\}\right], \qquad f_{c,\theta} = k_{c,\theta} f_{c'} \tag{1}$$

where the strain $\varepsilon_{cu,\theta}$ corresponds to stress $f_{c,\theta}$, the reduction factor can be chosen according to standard [22]. The reduction factors $k_{c,\theta}$ ($k_{c,\theta}$ = 0.925 for θ_c = 150°C) for the stress–strain relationship are considered in accordance with the standard.

The stress-strain relationships for steel (**Figure 4**) are considered in accordance with Eurocode [22] on dependency of temperature level for heating rates between 2 and 50 K/min. In the case of steel, the stress-strain diagram is divided into four regions.

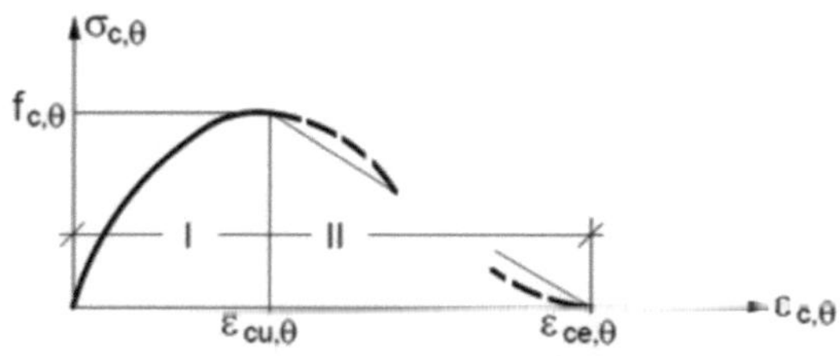

Figure 3. Stress-strain relationship of the concrete dependent on temperature [22].

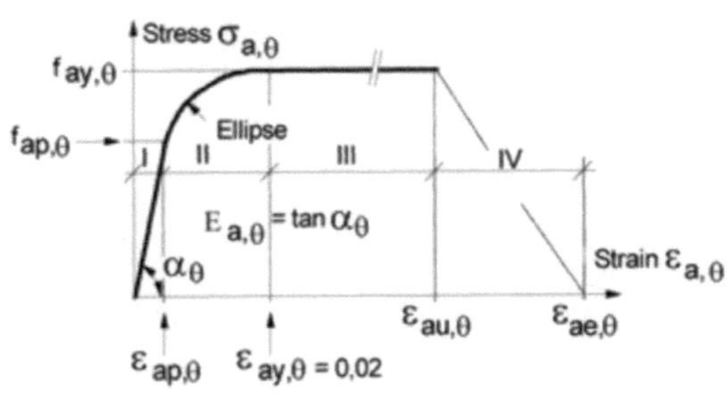

Figure 4. Stress-strain relationship of the steel dependent on temperature [22].

The stress–strain relations $\sigma_{a,\theta} \approx \varepsilon_{a,\theta}$ are defined in the following form in region I:

$$\sigma_{a,\theta} = E_{a,\theta}\varepsilon_{a,\theta}, \quad E_{a,\theta} = k_{E,\theta}E_a \tag{2}$$

where the reduction factor $k_{E,\theta}$ can be chosen according to the values of [10].

In region II:

$$\begin{aligned} &\sigma_{a,\theta} = (f_{ay} - c) + \frac{b}{a}\sqrt{a^2 - (\varepsilon_{ay,\theta} - \varepsilon_{a,\theta})^2}, a^2 = (\varepsilon_{ay,\theta} - \varepsilon_{ap,\theta})(\varepsilon_{ay,\theta} - \varepsilon_{ap,\theta} + c/E_{a,\theta}), \\ &b^2 = E_{a,\theta}(\varepsilon_{ay,\theta} - \varepsilon_{ap,\theta})c + c^2, c = \frac{(f_{ay,\theta} - f_{ap,\theta})^2}{E_{a,\theta}(\varepsilon_{ay,\theta} - \varepsilon_{ap,\theta}) - 2(f_{ay,\theta} - f_{ap,\theta})} \end{aligned} \tag{3}$$

and in region III:

$$\sigma_{a,\theta} = f_{ay,\theta} \tag{4}$$

A graphical display of the stress-strain relationships for steel grade S235 is presented in **Figure 4** up to the maximum strain of $\varepsilon_{ay,\theta} = 2\%$.

The strength and deformation properties of reinforcing steels under elevating temperatures may be obtained by the same mathematical model as that presented for structural steel S235. The reduction factors $k_{E,\theta}(k_{E,\theta} = 0.95$ for $\theta_a = 150°C)$ for the stress-strain relationship are considered in accordance with the standard.

The material properties of the concrete structures in the numerical model were considered using the experimental tests statistically evaluated during the performance of the nuclear power plant [3, 28]. The material properties of the steel structures were not changed during plant performance [3].

3.3. Nonlinear model of steel and reinforced concrete structures

The theory of large strain and rate-independent plasticity was proposed during the high-overpressure loading using the SHELL181-layered shell element from (**Figure 5**) the ANSYS library [26].

The vector of the displacement of the lth-shell layer $\{u^l\} = \{u^l_x, u^l_y, u^l_z\}^T$ is approximated by the quadratic polynomial [26] in the form

$$\{u^l\} = \begin{Bmatrix} u^l_x \\ u^l_y \\ u^l_z \end{Bmatrix} = \sum_{i=1}^{4} N_i.\begin{Bmatrix} u_{xi} \\ u_{yi} \\ u_{zi} \end{Bmatrix} + \sum_{i=1}^{4} N_i.\frac{\zeta.t_i}{2}.\begin{bmatrix} a_{1,i} & b_{1,i} \\ a_{2,i} & b_{2,i} \\ a_{3,i} & b_{3,i} \end{bmatrix}.\begin{Bmatrix} \theta_{xi} \\ \theta_{yi} \end{Bmatrix} \tag{5}$$

where N_i is the shape function for i-node of the 4-node shell element, u_{xi}, u_{yi}, and u_{zi} are the motions of i-node, ζ is the thickness coordinate, t_i is the thickness at i-node, $\{a\}$ is the unit vector in x-direction, $\{b\}$ is the unit vector in the plane of element and normal to $\{a\}$, θ_{xi} or θ_{yi} are the rotations of i-node about vector $\{a\}$ or $\{b\}$.

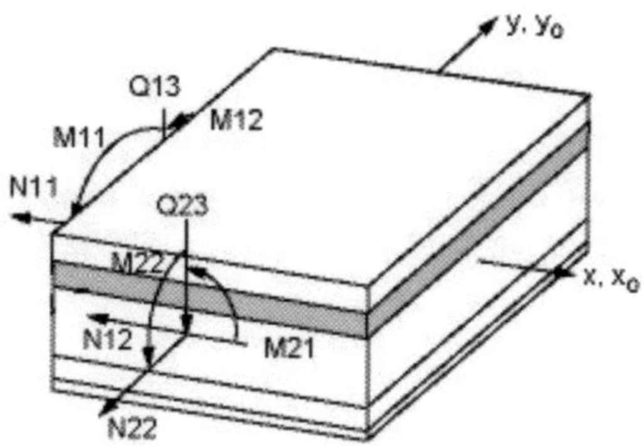

Figure 5. SHELL181-layered element with smeared reinforcements [26].

In the case of the elastic state, the stress–strain relations for the lth-layer are defined in the form

$$\{\sigma^l\} = [D_e^l]\{\varepsilon^l\} \tag{6}$$

where strain and stress vectors are as follows: $\{\varepsilon^l\}^T = \{\varepsilon_x, \varepsilon_y, \gamma_{xy}, \gamma_{yz}, \gamma_{zx}\}$, $\{\sigma^l\}^T = \{\sigma_x, \sigma_y, \tau_{xy}, \tau_{yz}, \tau_{zx}\}$ and the matrix of the material stiffness.

3.3.1. *Geometric nonlinearity*

If the rotations are large while the mechanical strains (those that cause stresses) are small, then it is possible to use a large rotation procedure. A large rotation analysis is performed in a static analysis in the ANSYS program [26].

From the following relations, the strain in the n-step of the solution can be computed:

$$\{\varepsilon_n\} = [B_o][T_n]\{u_n\} \tag{7}$$

Where $\{u_n\}$ is the displacement vector, $[B_o]$ is the original strain–displacement relationship and $[T_n]$ is the orthogonal transformation relating the original element coordinates to the convected (or rotated) element coordinates.

3.3.2. *Material nonlinearity*

The technology segments on board of the hermetic zone are made from the steel. The finite element model (FEM) of these segments is based on the HMH-yield criterion for the isotropic and homogenous material properties.

Consequently the stress–strain relations are obtained from the following relations

$$\{d\sigma\} = [D_{el}](\{d\varepsilon\} - \{d\varepsilon^{pl}\}) = [D_{el}]\left(\{d\varepsilon\} - d\lambda\left\{\frac{\partial Q}{\partial \sigma}\right\}\right) = [D_{ep}]\{d\varepsilon\} \tag{8}$$

where $[D_{ep}]$ is elastic-plastic matrix in the form

$$[D_{ep}] = [D_e] - \frac{[D_e]\left\{\frac{\partial Q}{\partial \sigma}\right\}\left\{\frac{\partial F}{\partial \sigma}\right\}^T [D_e]}{A + \left\{\frac{\partial F}{\partial \sigma}\right\}^T [D_e]\left\{\frac{\partial Q}{\partial \sigma}\right\}} \tag{9}$$

The hardening parameter A depends on the yield function and model of hardening (isotropic or kinematic). Huber-Mises-Hencky (HMH) yield is defined in the form

$$\sigma_{eq} = \sigma_T(\kappa), \tag{10}$$

Where σ_{eq} is the equivalent stress in the point and $\sigma_o(\kappa)$ is the yield stress that depends on the hardening.

In the case of kinematic hardening by Prager (vs Ziegler) and the ideal Bauschinger's effect, it is given as

$$A = \frac{2}{9E}\sigma_T^2 H' \tag{11}$$

The hardening modulus H' for this material is defined in the form

$$H' = \frac{d\sigma_{eq}}{d\varepsilon_{eq}^p} = \frac{d\sigma_T}{d\varepsilon_{eq}^p} \tag{12}$$

When this criterion is used with the isotropic hardening option, the yield function is given by

$$F(\sigma) = \sqrt{\{\sigma\}^T [M]\{\sigma\}} - \sigma_o(\varepsilon_{ep}) = 0 \tag{13}$$

where $\sigma_o(\varepsilon_{ep})$ is the reference yield stress, ε_{ep} is the equivalent plastic strain and the matrix [M] is as follows:

$$[M] = \begin{bmatrix} 1 & 0 & 0 & 0 & 0 & 0 \\ 0 & 1 & 0 & 0 & 0 & 0 \\ 0 & 0 & 1 & 0 & 0 & 0 \\ 0 & 0 & 0 & 2 & 0 & 0 \\ 0 & 0 & 0 & 0 & 2 & 0 \\ 0 & 0 & 0 & 0 & 0 & 2 \end{bmatrix} \tag{14}$$

3.3.3. Nonlinear material model of the concrete structures

The presented constitutive model is a further extension of the smeared and oriented crack model, which was developed in [3]. A new concrete cracking-layered finite shell element was developed and incorporated into the ANSYS system [3] considering the experimental tests of real reinforced concrete plate and wall structures. The layered shell elements are proposed considering the nonlinear properties of the concrete and steel depending on temperature.

The concrete compressive stress f_c, the concrete tensile stress f_t and the shear modulus G are reduced during the crushing or cracking of the concrete. These effects are updated on the numerical model.

http://dx.doi.org/10.5772/intechopen.68772

In this model, the stress-strain relation is defined (**Figure 6**) following CEB-FIP Model Code [25]:

- Loading-compression region $\varepsilon_{cu} < \varepsilon^{eq} < 0$

$$\sigma_c^{ef} = f_c^{ef} \cdot \frac{k.\eta - \eta^2}{1 + (k-2).\eta}, \eta = \frac{\varepsilon^{eq}}{\varepsilon_c}, (\varepsilon_c \doteq -0.0022, \varepsilon_{cu} \doteq -0.0035) \tag{15}$$

- Softening-compression region $\varepsilon_{cm} < \varepsilon^{eq} < \varepsilon_{cu}$

$$\sigma_c^{ef} = f_c^{ef} \cdot \left(1 - \frac{\varepsilon^{eq} - \varepsilon_c}{\varepsilon_{cm} - \varepsilon_{cu}}\right) \tag{16}$$

- The tension region $\varepsilon_t < \varepsilon^{eq} < \varepsilon_m$

$$\sigma_c^{ef} = f_t \cdot \exp\left(-2.(\varepsilon^{eq} - \varepsilon_t)/\varepsilon_{tm}\right), (\varepsilon_t \doteq 0.0001, \varepsilon_{tm} \doteq 0.002) \tag{17}$$

The equivalent values of f_t^{eq} and f_c^{eq} were considered for the plane stress state. The relation between the one and bidimensional stresses state was considered in the plane of principal stresses (σ_{c1}, σ_{c2}) of each shell layer by Kupfer (see **Figure 7**) [3].

The shear concrete modulus G was defined for cracking concrete by Kolmar [23] in the form

$$G = r_g.G_o, r_g = \frac{1}{c_2}\ln\left(\frac{\varepsilon_u}{c_1}\right), c_1 = 7 + 333(p - 0.005), c_2 = 10 - 167(p - 0.005) \tag{18}$$

where G_o is the initial shear modulus of concrete, ε_u is the strain in the normal direction to crack, c_1 and c_2 are the constants dependent on the ratio of reinforcing and p is the ratio of reinforcing transformed to the plane of the crack ($0 < p < 0.02$).

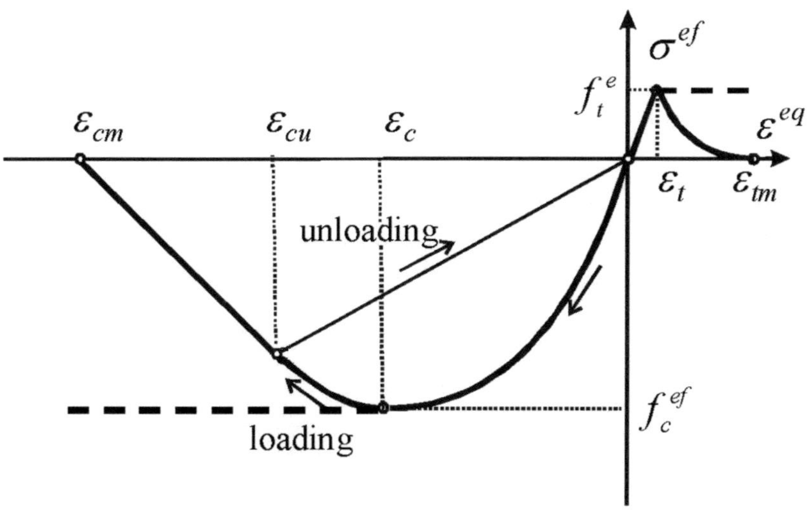

Figure 6. The concrete stress-strain diagram.

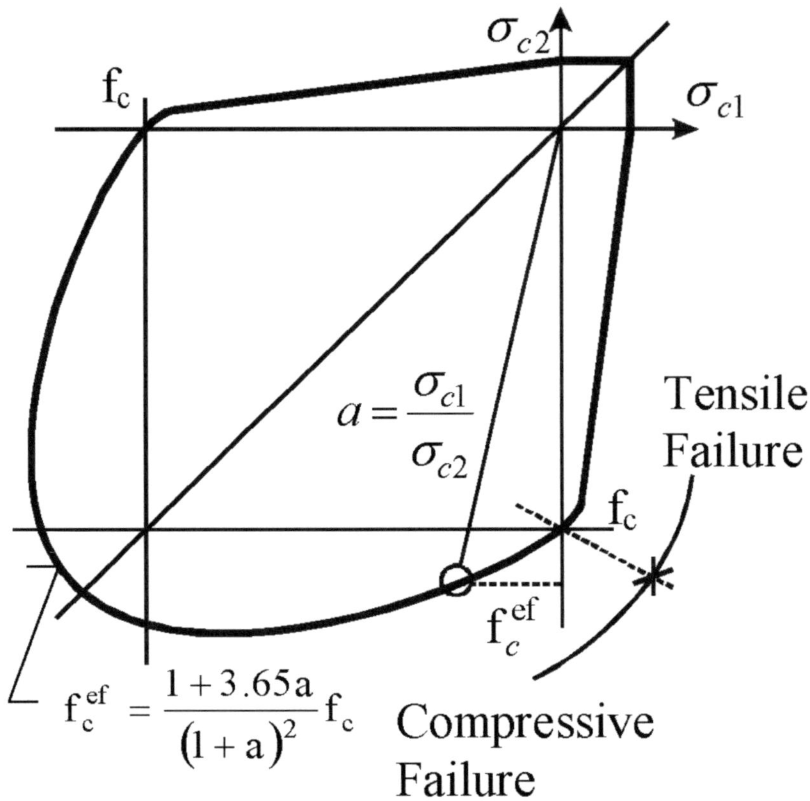

Figure 7. Kupfer's plasticity function.

The strain-stress relationship in the Cartesian coordinates can be defined in dependency on the direction of the crack (in the direction of principal stress, vs strain)

$$[\sigma_{cr}] = [D_{cr}]\{\varepsilon_{cr}\} \text{ and } [\sigma] = [T_\sigma]^T[D_{cr}][T_\varepsilon]\{\varepsilon\} \tag{19}$$

$$[D^l_{cr}] = [T^l_{c.\sigma}]^T[D^l_{cr}][T^l_{c.\varepsilon}] + \sum_{s=1}^{N_{rein}}[T^l_s]^T[D^l_s][T^l_s] \tag{20}$$

where $[T_{c.\sigma}]$, $[T_{c.\varepsilon}]$ and $[T_s]$ are the transformation matrices for the concrete and the reinforcement separately, N_{rein} is the number of the reinforcements in the lth–layer (**Figure 8**).

The stress-strain relationship for the concrete lth-layer cracked in one direction is

$$\begin{Bmatrix} \sigma_1 \\ \sigma_2 \\ \tau_{12} \\ \tau_{13} \\ \tau_{23} \end{Bmatrix}_l = \begin{bmatrix} 0 & 0 & 0 & 0 & 0 \\ 0 & E & 0 & 0 & 0 \\ 0 & 0 & G^{cr}_{12} & 0 & 0 \\ 0 & 0 & 0 & G^{cr}_{13} & 0 \\ 0 & 0 & 0 & 0 & G^{cr}_{23} \end{bmatrix}_l \begin{Bmatrix} \varepsilon_1 \\ \varepsilon_2 \\ \gamma_{12} \\ \gamma_{13} \\ \gamma_{23} \end{Bmatrix}_l \tag{21}$$

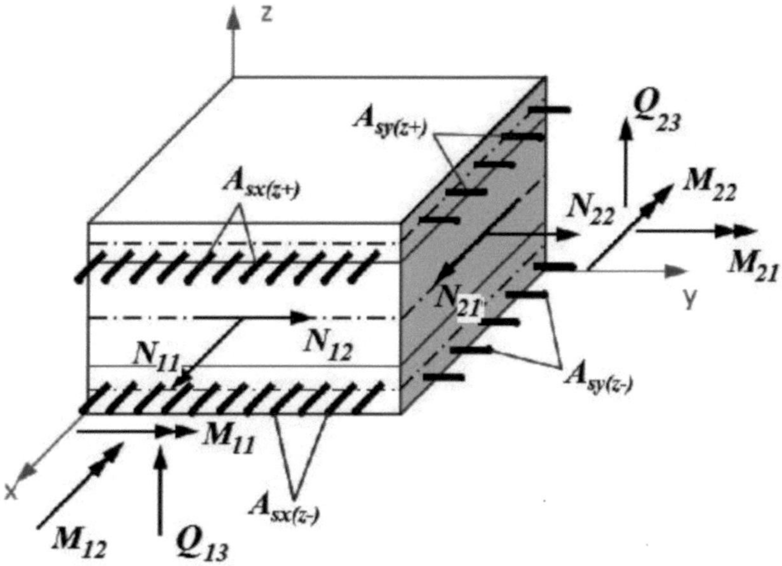

Figure 8. SHELL181-layered element with smeared reinforcements.

When the tensile stress in the 2-directions reaches the value f'_t, the latter cracked plane perpendicular to the first one is assumed to be formed, and the stress–strain relationship becomes

$$\begin{Bmatrix} \sigma_1 \\ \sigma_2 \\ \tau_{12} \\ \tau_{13} \\ \tau_{23} \end{Bmatrix}_l = \begin{bmatrix} 0 & 0 & 0 & 0 & 0 \\ 0 & 0 & 0 & 0 & 0 \\ 0 & 0 & G_{12}^{cr}/2 & 0 & 0 \\ 0 & 0 & 0 & G_{13}^{cr} & 0 \\ 0 & 0 & 0 & 0 & G_{23}^{cr} \end{bmatrix}_l \begin{Bmatrix} \varepsilon_1 \\ \varepsilon_2 \\ \gamma_{12} \\ \gamma_{13} \\ \gamma_{23} \end{Bmatrix}_l \tag{22}$$

where the shear moduli are reduced by parameters r_{g1} and r_{g2} by Kolmar [27] as follows:

$$G_{12}^{cr} = G_o.r_{g1}, G_{13}^{cr} = G_o.r_{g1}, G_{23}^{cr} = G_o r_{g2}.$$

The stress-strain relationship defined in the direction of the principal stresses must be transformed to the reference axes *XY*. The simplified smeared model of the concrete cracked is more convenient for finite element formulation than the singular discrete model.

The smeared calculation model is determined by the size of the finite element, hence its characteristic dimension $L_c = \sqrt{A}$, where A is the element area (vs integrated point area of the element). The assumption of constant failure energies G_f = const is proposed in the form

$$G_f = \int_0^{\infty} \sigma_n(w) dw = A_G.L_c, \quad w_c = e_w.L_c \tag{23}$$

where w_c is the width of the failure, σ_n is the stress in the concrete in the normal direction and A_G is the area under the stress-strain diagram of concrete in tension. The descend line of concrete stress-strain diagram can be defined on dependency on the failure energies [25] by the modulus in the form

$$E_{c,s} = E_c/(1-\lambda_c), \qquad \lambda_c = 2G_f E_c/(L_c.\sigma_{\max}^2) \tag{24}$$

where E_c is the initial concrete modulus elasticity, $\sigma_{\max}$ is the maximal stress in the concrete tension. From the condition of the real solution of relation (18), it follows that the characteristic dimension of element must satisfy the following condition:

$$L_c \leq 2G_f E_c/\sigma_{\max}^2 \tag{25}$$

The characteristic dimension of the element is determined by the size of the failure energy of the element. The theory of a concrete failure was implied and applied to the two-dimensional (2D)-layered shell elements SHELL181 in the ANSYS element library [26]. The CEB-FIP Model Code [25] defines the failure energies G_f [N/mm] depending on the concrete grades and the aggregate size d_a as follows:

$$G_f = (0.0469d_a^2 - 0.5d_a + 26)(f_c/10)^{0,7} \tag{26}$$

The limit of damage at a point is controlled by the values of the so-called crushing or total failure function F_u. The modified Kupfer's condition for the lth-layer of section is as follows:

$$F_u^l = F_u^l(I_{\varepsilon 1}; I_{\varepsilon 2}; \varepsilon_u) = 0, \quad F_u^l = \sqrt{\beta(3J_{\varepsilon 2} + \alpha I_{\varepsilon 1})} - \varepsilon_u = 0\,, \tag{27}$$

where $I_{\varepsilon 1}$, $I_{\varepsilon 2}$ are the strain invariants, and ε_u is the ultimate total strain extrapolated from uniaxial test results (ε_u = 0.002 in the tension domain, or ε_u = 0.0035 in the compression domain), and α, β are the material parameters determined from Kupfer's experiment results (β = 1.355, α = 1.355ε_u).

The failure function of the whole section will be obtained by the integration of the failure function through the whole section in the form

$$F_u = \frac{1}{t}.\int_0^t F_u^l(I_{\varepsilon 1}; I_{\varepsilon 2}; \varepsilon_u)\, dz = \frac{1}{t}\sum_{l=1}^{N_{lay}} F_u^l(I_{\varepsilon 1}; I_{\varepsilon 2}; \varepsilon_u) t_l \tag{28}$$

where t_l is the thickness of the lth-shell layer, t is the total shell thickness and N_{lay} is the number of layers.

The maximum strain ε_s of the reinforcement steel in the tension area ($\max(\varepsilon_s) \leq \varepsilon_{sm} = 0.01$) and by the maximum concrete crack width w_c ($\max(w_c) \leq w_{cm} = 0.3$ mm) determine the local collapse of reinforced concrete structure.

The program CRACK based on the presented nonlinear theory of the layered reinforced concrete shell was adopted in the software ANSYS [3]. These procedures were tested in comparison with the experimental results [3, 15, 17, 28].

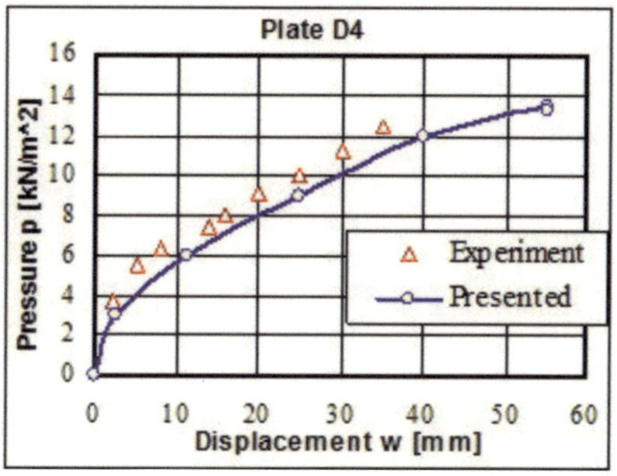

Figure 9. Comparison of experimental and nonlinear numerical analysis of plates.

The reinforced concrete plates D4 (**Figure 9**) with the dimensions 3590/1190/120 mm were simply supported and loaded by pressure p on the area of plate. The plate D4 was reinforced by steel grid KARI Ø8 mm, a′ 150 × 150 mm at the bottom. Material characteristics of plate D4 are the following: Concrete, E_c = 30.9 Gpa, μ = 0.2, f_c = −34.48 Mpa, f_t = 4.5 Mpa and the Reinforcement, E_s = 210.7 Gpa, μ = 0.3, f_s = 550.3 Mpa.

4. Probabilistic assessment

Recent advances and the general accessibility of information technologies and computing techniques give rise to assumptions concerning the wider use of the probabilistic assessment of the reliability of structures through the use of simulation methods in the world [3, 14–22, 29–40]. The probabilistic definition of the reliability condition is of the form

$$g(R, E) = R - E \geq 0 \tag{29}$$

where $g(R,E)$ is the reliability function.

In the case of simulation methods, the failure probability is defined as the best estimation on the base of numerical simulations in the form

$$p_f = \frac{1}{N} \sum_{i=1}^{N} I[g(X_i) \leq 0] \tag{30}$$

where N in the number of simulations, $g(.)$ is the failure function and $I[.]$ is the function with value 1, if the condition in the square bracket is fulfilled, otherwise is equal to 0. The semi- or full-probabilistic methods can be used for the estimation of the structure failure in the critical structural areas. In the case of the semi-probabilistic method, the probabilistic simulation in the critical areas is based on the results of the nonlinear analysis of the full FEM model for the median values of the input data. The full probabilistic method result from the nonlinear analysis of the series simulated cases considered the uncertainties of the input data.

4.1. Uncertainties of the input data

The action effect E and the resistance R are calculated considering the uncertainties of the input data as follows [3]:

$$E = G_k g_{\text{var}} + Q_k q_{\text{var}} + P_k p_{\text{var}} + T_k t_{\text{var}} \quad \text{and} \quad R = R_k r_{\text{var}} \tag{31}$$

The uncertainties of the input data were taken in accordance with the standard requirements [29, 30–32] (**Table 3**).

4.2. Probabilistic simulation methods

Various simulation methods (direct, modified or approximation methods) can be used for the consideration of the influences of the uncertainty of the input data.

In the case of the nonlinear analysis of the full FEM model, the approximation method RSM (response surface method) is the most effective method [3].

The RSM method is based on the assumption that it is possible to define the dependency between the variable input and the output data through the approximation functions in the following form:

$$Y = c_o + \sum_{i=1}^{N} c_i X_i + \sum_{i=1}^{N} c_{ii} X_i^2 + \sum_{i=1}^{N-1} \sum_{j>i}^{N} c_{ij} X_i X_j \tag{32}$$

where c_o is the constant member; c_i are the constants of the linear member and c_{ij} for the quadratic member, which are given for predetermined schemes for the optimal distribution of the variables or for using the regression analysis after calculating the response. The 'Central

	Quantities		Histograms		
Input data	Charact. value	Variable value	Type	Mean	Deviation
				μ	σ [%]
Dead load	G_k	g_{var}	N	1	10
Live load	Q_k	q_{var}	Beta	0.643	22.6
Pressure	P_k	p_{var}	N	1	8
Temperature	T_k	t_{var}	Beta	0.933	14.1
Model	E_k	e_{var}	N	1	5
Resistance	R_k	r_{var}	N	1	5

Table 3. The variability of input parameters.

Composite Design Sampling' (CCD) method or the 'Box-Behnken Matrix Sampling' (BBM) method [3] can be used to determine the polynomial coefficients.

The philosophy of the RSM method is presented in **Figure 10**. The original system of the global surface is discretized using approximation function. The design of the experiment determines the polynomic coefficients. The efficiency of computation of the experimental design depends on the number of design points. With the increase of the number of random variables, this design approach becomes inefficient. The central composite design, developed by Box and Wilson, is more efficient.

The central CCD method is defined as follows (**Figure 10**):

1. Factorial portion of design—a complete 2k factorial design (equals −1, +1).
2. Centre point—no centre points, no ≥1 (generally no = 1).
3. Axial portion of design—two points on the axis of each design variable at distance α from the design centre.

The total number of design points is equal to $N = 2^k + 2k + n_o$. The sensitivity of the variables is determined by the correlation matrices. The RSM method generates the explicit performance function for the implicit or complicated limit-state function. This method is very effective to solve robust and complicated tasks.

4.3. Evaluation of the fragility curve

The PSA approach to the evaluation of probabilistic pressure capacity involves limit-state analyses. The limit states should represent possible failure modes of the confinement functions. The identification of potential failure modes is the first step of a probabilistic containment overpressure evaluation. For each failure mode, the median values were established based on the used failure criteria dependent on the applied loading consisting of temperature, pressure and dead load. Along with the pressure capacities for the leak-type failure modes, leak areas are to be estimated in a probabilistic manner. The expected leak areas are failure mode dependent. After calculation of the fragility or conditional probability of failure of containment at different locations, we must consider a combination of pressure-induced failure probabilities of different break or leak locations within containment.

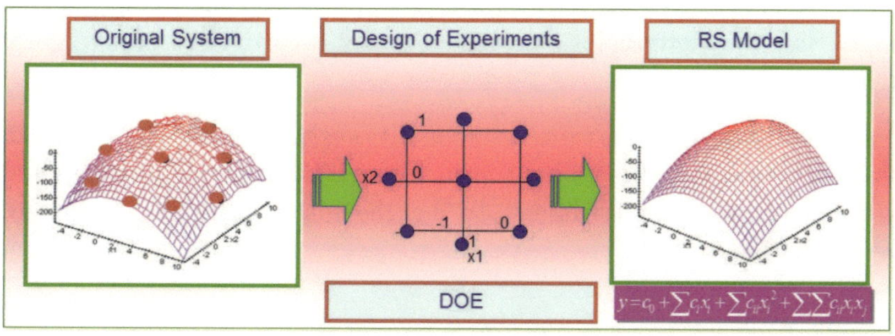

Figure 10. Scheme of the RSM approximation method with the CCD design experiment.

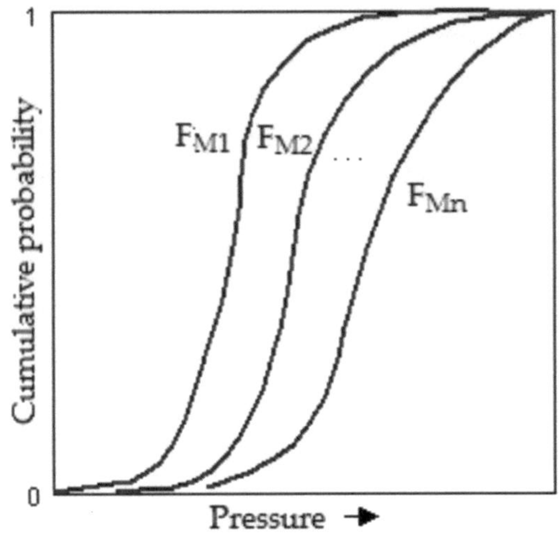

Figure 11. Family of fragility curves showing modelling uncertainty.

Containment may fail at different locations under different failure modes (see **Figure 11**). Consider two failure modes A and B, each with n fragility curves and respective probabilities p_i ($i = 1, \ldots, n$) and q_j ($j = 1, \ldots, n$). Then, the union $C = A \cup B$, the fragility $F_{Cij}(x)$ is given by

$$F_{Cij}(x) = F_{Ai}(x) + F_{Bj}(x) - F_{Ai}(x) \cap F_{Bj}(x) \tag{33}$$

where the subscripts i and j indicate one of the n fragility curves for the failure modes and x denotes a specific value of the pressure within the containment. The probability p_{ij} associated with fragility curve $F_{Cij}(x)$ is given by p_i. q_j if the median capacities of the failure modes are independent. The result of the intersection term in Eq. (32) is $F_{Aj}(x).F_{Bj}(x)$ when the randomness in the failure mode capacities is independent and min $[F_{Ai}(x), F_{Bj}(x)]$ when the failure modes are perfectly dependent.

The flow is the consequence of an accident that depends on the total leak area. Multiple leaks at different locations of the containment (e.g. bellows, hatch and airlock) may contribute to the total leak area. Using the methodology described earlier, we can obtain the fragility curves for leak at each location.

For a given accident sequence, the induced accident pressure probability distribution, $h(x)$, is known. This is convolved with the fragility curve for each leak location to obtain the probability of leak from that location (P_{Li}). It is understood that there is no break or containment rupture at this pressure.

$$p_{Li} = \int_0^\infty h(x)[1 - F_b(x)]F_l(x)dx \tag{34}$$

Here, the $F_b(x)$ is the fragility of break at the location and $F_l(x)$ is the fragility of the leak. The leak is for each location specified as a random variable with a probability distribution.

5. Probabilistic analysis of the concrete containment

The NPP buildings with the reactor VVER 440/213 consist of the turbine hall, middle building, reactor building and bubble condenser [3]. The building of the power block was idealized with a FEM model consisting of 28 068 elements with 104 287 DOF (**Figures 12** and **13**).

The roof plate of the bubbler tower was defined as the critical area of the containment failure. The probability of the containment failure was determined for various levels of the overpressures Δp = 250, 300, 350, 400, 450 kPa. The probability of the failure was determined by Eqs. (21) and (22) for 10^6 Monte Carlo simulations in program FReET [32].

5.1. Nonlinear deterministic analysis

On the basis of the nonlinear analysis due to the monotone increasing of overpressure inside the hermetic zone, the critical sections of the structure were determined [3]. The resistance of these critical sections was analysed taking into account the design values of the material characteristics and the load. The slab at the top of the bubbler tower building was defined as

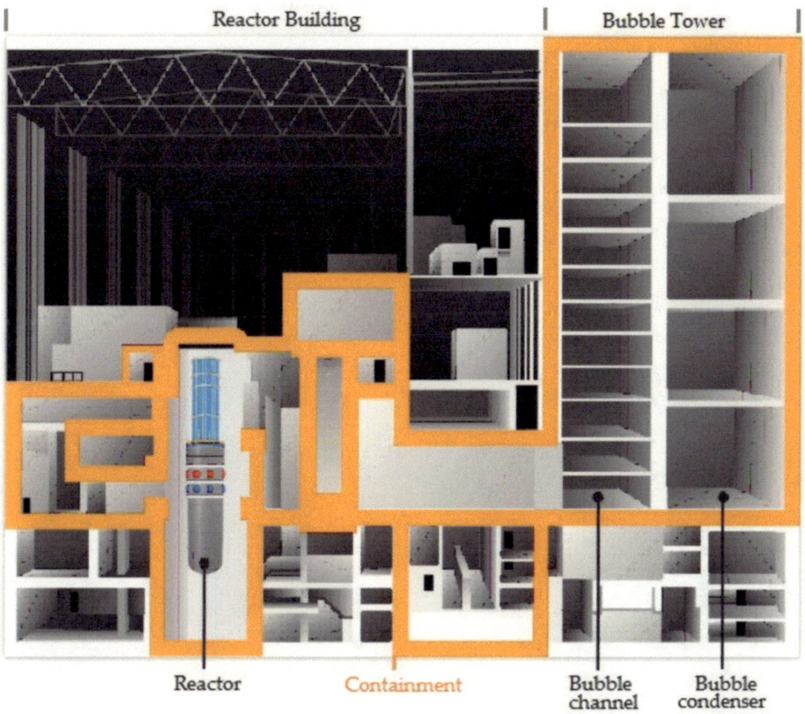

Figure 12. Section plane of the condenser containment.

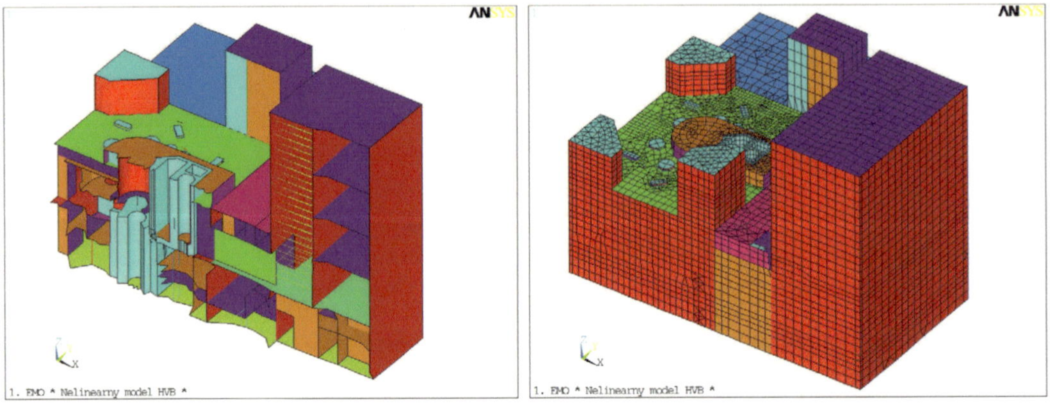

Figure 13. Calculation model of NPP VVER 440/213.

the critical area of hermetic zone failure. The cracking process started at the concrete slab along the middle wall due to concentration of the temperature and overpressure effects. There is the effective temperature gradient equal to 60–90°C in the middle plane of the wall and the plate. The mean value of the critical overpressure was equal to 352.5 kPa, and the max. strain is lower than 0.002 in the middle plane of the reinforced concrete panel.

The cracking process ($\varepsilon_1 \geq \varepsilon_t \doteq 0.0001$) at the bottom or top section of the reinforced concrete panels started when the overpressure was equal to 250 kPa.

The interior structures of the hermetic zone are loaded with the accident temperature equal to 150°C and the outside structures in the contact with the exterior are loaded by −28°C. The difference between the interior end of the exterior temperatures has significant influences to the peak strain in the structures.

The comparison of the strain shape from the linear and nonlinear solutions is presented in **Figure 14** and the stress shape in **Figure 15**. The strain increases and the stress decreases in the nonlinear solution in comparison with the linear solution.

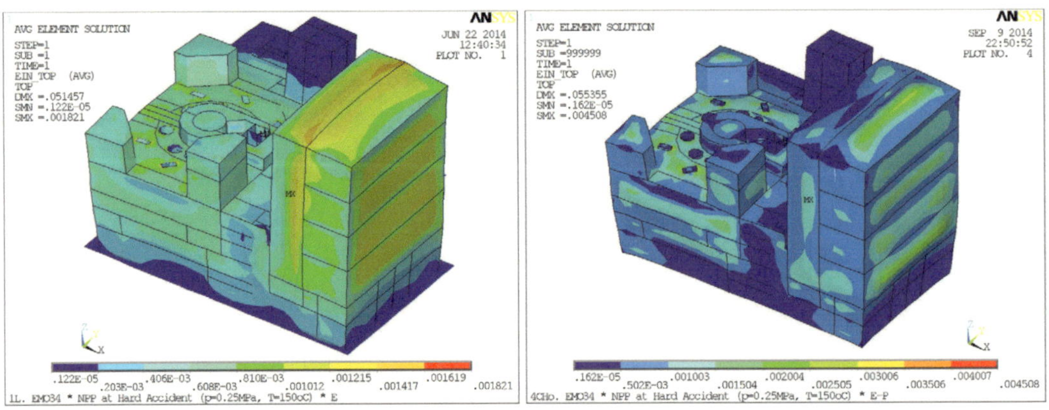

Figure 14. Strain intensity from the linear and nonlinear analysis.

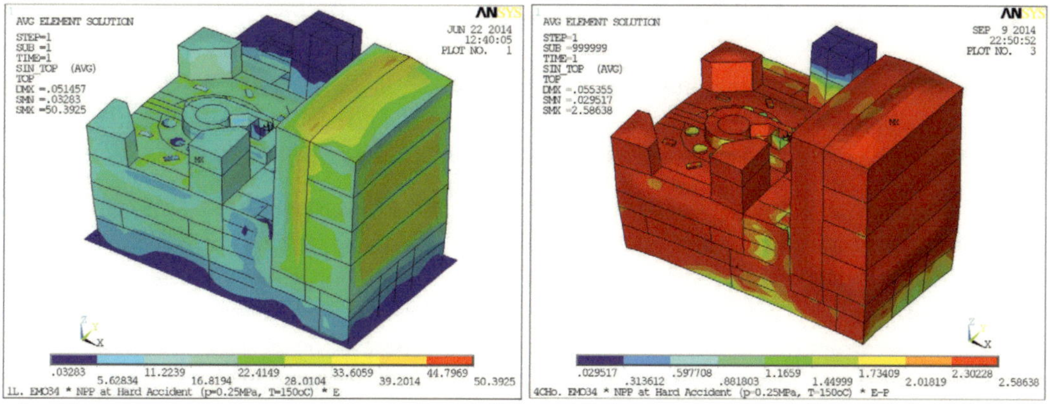

Figure 15. Stress intensity from the linear and nonlinear analysis.

5.2. Evaluation of the fragility curve

The probability of the containment failure was determined for this critical structure on the basis of the nonlinear deterministic analysis of the containment for various levels of the overpressure. The function of the failure was considered by Eqs. (28) and (29) for 10^6 Monte Carlo simulations in program FReET [32]. The probability of containment failure (**Figure 16**) is calculated from the probability of the reliability function *RF* in the form

$$P_f = P(RF < 0) \text{ and } RF = 1 - F_u(I_{\varepsilon 1}; J_{\varepsilon 2}; \varepsilon_u)/\varepsilon_u, \quad (35)$$

where the reliability condition *RF* is defined depending on a concrete failure condition (30).

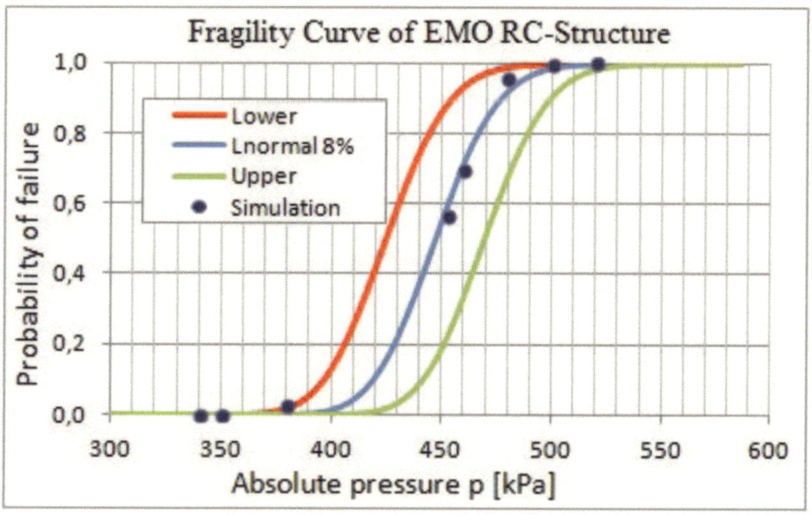

Figure 16. Fragility curve of failure pressure in the critical area of the hermetic zone under the accidental internal temperature $T_i = 150°C$ and external temperature $T_e = -28°C$.

6. Probabilistic analysis of the steel technology segments

The reactor and the bubble condenser-reinforced structures with steel liner represent the critical structures of the NPP hermetic zone [3]. Among the critical technology structures, the reactor hermetic covers and the reactor-protective hood, belong too [39, 40].

The reactor-protective hood is shown in **Figure 17**. The protective hood is an all-welded structure consisting of a spherical and a cylindrical part. The spherical part has a manhole of 500 mm in diameter with a ladder. The manhole facilitates equipment maintenance in the concrete cavity without the necessity to remove the protective hood. In order to ensure higher strength of the structure (on a seismic event), the protective hood is reinforced with a pipe (inner Ø712 mm) and six ribs. At the top, the pipe is welded in the canter of the hood spherical part, while the other end covers the ring on the upper block beam. The protective hood is set on a counter-flange and is attached to it with 60 M48 bolts and sealed with packing. The cap structure includes a platform with railing.

The finite element model of reactor cover was created in software ANSYS by the shell, beam, combine and mass elements. The envelope of cover is from layered shell elements (SHELL181). The surface load is defined using three-dimensional (3D) structural surface elements (SURF154). The connection with bolts is modelled by the combine elements (COMBINE14). The element of point mass (MASS21) represents the concentrated masses adequate to local load of the technology, beam elements (BEAM44) for frame and beam connection. The contact element (CE) and links (CP) were used for the joint connection. The upper part of the hood has

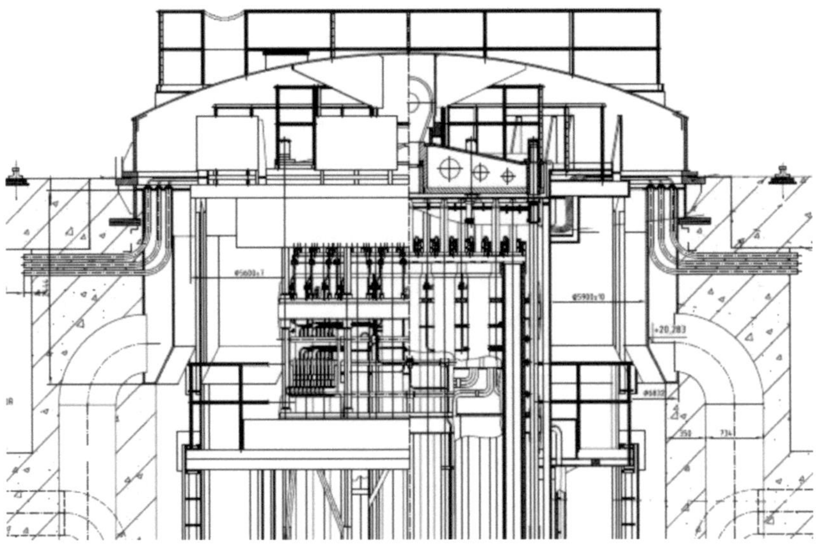

Figure 17. Vertical section of the reactor with reactor-protective hood.

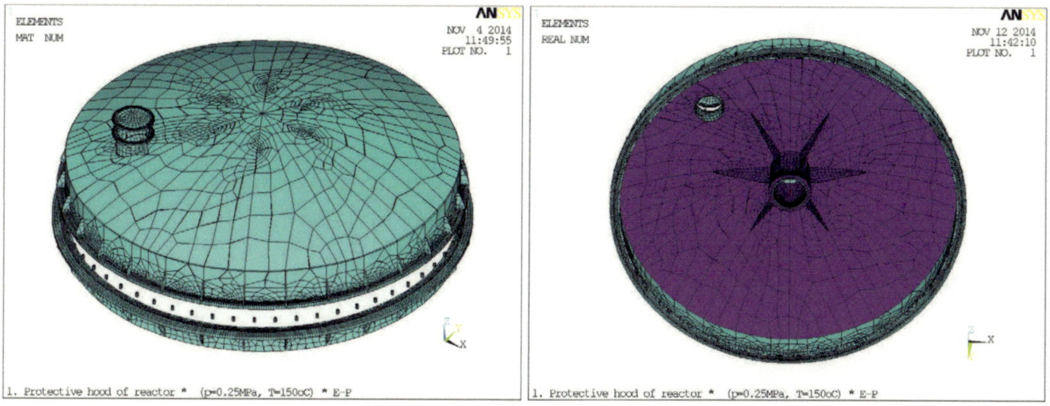

Figure 18. FEM model (RCOV) of the reactor-protective hood.

lugs used to handle it. The FEM model (RCOV) consists of 27,824 shell, beam and mass elements with 22,887 nodes (**Figure 18**).

6.1. Probability and sensitivity nonlinear analysis

The calculation of the probability of the reactor cover failure is based on the results of the nonlinear analysis for various levels of the accident pressure and mean values of the material properties. The critical areas of the technology segments defined from the nonlinear deterministic analysis are the mechanical closures. The CCD method of the RSM approximation is based on 45 nonlinear simulations depending on the six variable input data. The nonlinear solution for a single simulation consists of about 50–150 iterations depending on the scope of the plastic deformations in the calculated structures. The sensitivity analysis gives us the information about the influences of the variable properties of the input data on the output data (see **Figures 19** and **20**). These analyses are based on the correlation matrixes.

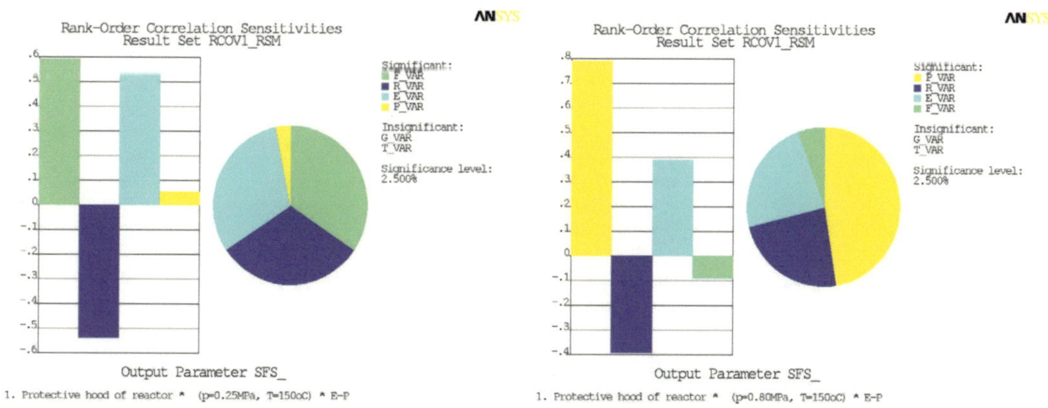

Figure 19. Sensitivity analysis of the safety function of the reactor-protective hood for overpressure Δp = 0.25 MPa and Δp = 0.80 MPa.

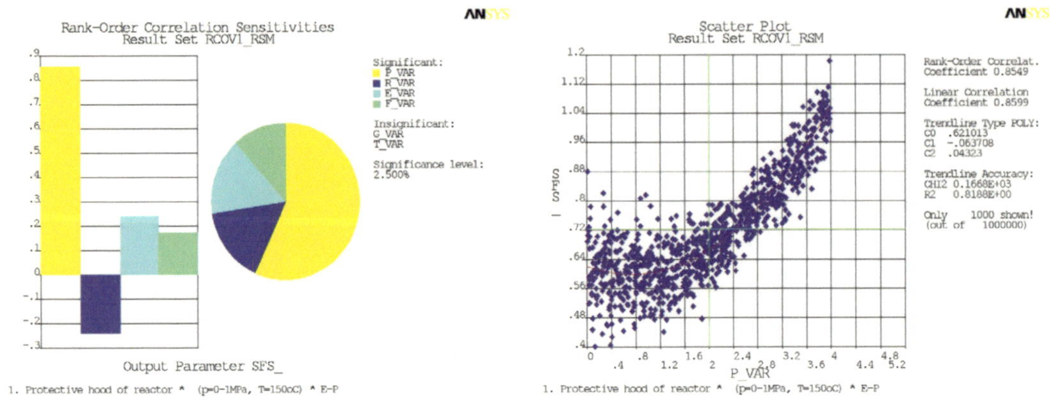

Figure 20. Sensitivity and trend analysis of the safety function of the reactor cover for uniform distribution of overpressure.

6.2. Fragility curves of failure pressure

The fragility curve of the failure pressure (see **Figure 21**) was determined performing 45 probabilistic simulations using the RSM approximation method with the experimental design CCD for 10^6 Monte Carlo simulations for each model and five levels of the overpressure. Various probabilistic calculations for five constant levels of overpressure next for the variable overpressure for gauss and uniform distribution were taken out. The nonlinear analysis of the steel technology structures was performed considering HMH-plastic criterion with the multilinear kinematic hardening stress-strain relations for various levels of the temperatures and the degradation of the strength. The uncertainties of the input data (**Table 3**) were

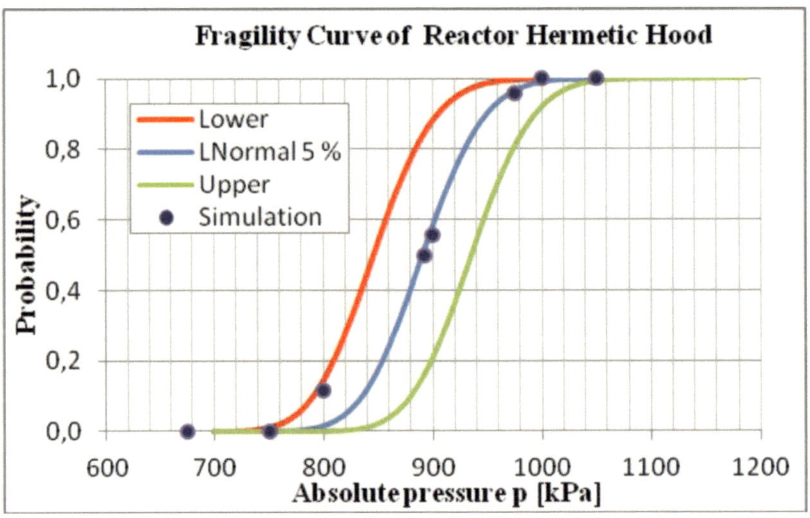

Figure 21. Fragility curves of the reactor-protective hood determined analytically for normal distribution with 5% envelope.

considered in accordance with the JCSS standards [30], NRC [11] and IAEA requirements [9]. The geometric and material nonlinearities of the steel solid and shell-layered elements were considered for the overpressure static loads from 250 to 1000 kPa.

7. Conclusions

The probability nonlinear analysis of the concrete containment failure was made for the overpressure loads from 250 to 500 kPa using the nonlinear solution of the reinforced concrete shell-layered elements. The CRACK program, which was developed by the author and implemented into the ANSYS system [3, 14–18], was used to perform the nonlinear analyses. The uncertainties of the load levels (temperature, dead and live loads), the material model of the composite structure (concrete cracking and crushing), reinforcement, and liner as well as other influences following from the inaccuracy of the calculated model and the numerical methods were taken into account in the Monte Carlo simulations [3]. The probability of the loss of the concrete containment integrity is less than 10^{-6} for the original structural model. The containment failure is equal to 0.050422906 for the overpressure of 275.5 kPa. The critical technology segment of the containment is the reactor-protective hood with the failure pressure $p_{u.0.05} = 766.9$ kPa. The mean value of pressure capacity of the reactor-protective hood is $p_{u.0.50} = 891.8$ kPa, the 95% upper bound is $p_{u.0.95} = 973.6$ kPa.

Acknowledgements

The project was performed with the financial support of the Grant Agency of the Slovak Republic (VEGA 1/0265/16) and the Slovak Republic within the Research and Development Operational Programme for the project 'University Science Park of STU Bratislava', ITMS 26240220084.

Author details

Juraj Králik

Address all correspondence to: juraj.kralik@stuba.sk

Faculty of Civil Engineering, Slovak University of Technology in Bratislava, Bratislava, Slovakia

References

[1] HSG26. Safety Assessment Principles for Nuclear Facilities. 0 ed. http://www.onr.org.uk/saps/. Office for Nuclear Regulation; 2014. p. 226. ISBN: 978 0 7176 6456 6

[2] Pioro I. Handbook of Generation IV Nuclear Reactors. 1st ed. www.elsevier.com/books/handbook-of-generation-iv-nuclear-reactors/pioro. Woodhead Publishing; 2016. p. 940. ISBN: 9780081001622

[3] Králik J. Safety and Reliability of Nuclear Power Buildings in Slovakia. Earthquake-Impact-Explosion. 1st ed. www.martinus.sk/?uItem=103269. Bratislava: STU Bratislava; 2009. p. 307. ISBN: 978-80-227-3112-6

[4] NUREG/CR-1150. Severe Accident Risks, An Assessment for Five US Nuclear Power Plants, Summary Report, Final Summary Report. 1st ed. www.nrc.gov/reading-rm/doc-collections/nuregs/staff/sr1150/v1/. U.S. NRC; 1990. p. 59

[5] ENSREG. Stress Tests Performed on European Nuclear Power Plants, Post Fukushima Accident. 1st ed. www.oecd.nea.org/nsd/fukushima/documents. Brussels: European Commission Brussels; 2012. p. 53. DOI: v12i-2012-04-25

[6] NEA/CNRA/R12. Proceedings of the Forum on the Fukushima Accident: Insights and Approaches. 1st ed. www.oecd-nea.org/nsd/docs/2011/cnra-r2011-12.pdf. OECD; 2011. p. 127. DOI: JT03312480

[7] NRA SR. The Stress Tests for Nuclear Power Plants in Slovakia. 1st ed. www.ujd.gov.sk/files/NS_Zatazove_testy_JE_SR_ang.pdf. UJD SR; 2011. p. 186

[8] IAEA, Safety Standards No. SSR-2/1, Safety of Nuclear Power Plants: Design, Specific Safety Requirements. 1st ed. www/.iaea.org/books. Vienna: IAEA Vienna; 2012. p. 91. ISSN 1020–525X

[9] IAEA. Safety Series No. SSG-4, Development and Application of Level 2 Probabilistic Safety Assessment for Nuclear Power Plants. www-pub.iaea.org/books/iaeabooks/8236/Development-and-Application-of-Level-2-Probabilistic-Safety-Assessment-for-Nuclear-Power-Plants-Specific-Safety-Guide. IAEA; 2010. p. 88. ISBN: 978-92-0-102210-3

[10] IAEA TECDOC No. 1791. Considerations on the Application of the IAEA Safety Requirements for the Design of Nuclear Power Plants. www-pub.iaea.org/books/iaeabooks/11069/Considerations-on-the-Application-of-the-IAEA-Safety-Requirements-for-the-Design-of-Nuclear-Power-Plants. Vienna: IAEA Vienna; 2016. p. 84. ISBN: 978-92-0-104116-6

[11] NRC, RG 1.200. An Approach for Determining the Technical Adequacy of Probabilistic Risk Assessment Results for Risk-Informed Activities. www.nrc.gov/docs/ML0904/ML090410014.pdf. Washington, DC: U.S. Nuclear Regulatory Commission; 2009

[12] NUREG/CR-6813. Issues and Recommendations for Advancement of PRA Technology In Risk-Informed Decision Making. www.nrc.gov/reading-rm/doc-collections/nuregs/contract/cr6813/. Washington, DC: U.S. NRC Washington; 2003. p. 60. DOI: NRC Job Code B1564

[13] Kim DH, Ahn KI. Comparison of Nuclear Power Plant Risk using Dose-Based Risk Measure, Nuclear Science and Technology. Vol. 1. 2011 . pp. 440–443. DOI: 10.15669/pnst.1.440

[14] Králik J. A RSM method for probabilistic nonlinear analysis of reinforced concrete bubbler tower structure integrity. In: Proceedings of European Safety and Reliability Conference, ESREL 2009, (Eds); 7–10 September, Vol. 2; Taylor & Francis Group, Prague, CR; 2009. pp. 1369–1372. ISBN: 978-0-415-55509-8

[15] Králik J. Probabilistic safety analysis of the nuclear power plants in Slovakia. Journal of KONBiN, Safety and Reliability Systems. 2010;**2,3 (14, 15)** (VERSITA Central European Science Publishers):35–48. ISSN 1895–8281

[16] Králik J, Králik J, Jr. Probabilistic nonlinear analysis of bubble tower structure due to extreme pressure and temperature. In: Applied Mechanics and Materials: Selected, Peer Reviewed Papers from the Engineering Mechanics 2015; May 2015, Vol. 821; Svratka, CR; 2016. 725–732. ISSN 1660–9336.

[17] Králik J et al. Structural reliability for containment of VVER 440/213 type. In: Nowakowski et al., editors. Safety and Reliability: Methodology and Applications. Taylor & Francis Group, London; 2015. pp. 2279–2286. ISBN: 978-1-138-02681-0

[18] Králik J. Experimental and numerical analysis of the hermetic tightness of NPP bubble tower structure. In: Proceedings of the 3rd International Scientific Conference, Structural and Physical Aspects of Construction Engineering, SPACE 2016; November 2016; High Tatras. ISBN: 978-80-553-2643-6.

[19] Králik J. Probabilistic nonlinear analysis of the hermetic circular cover of main coolant pump failure due to extreme pressure and temperature. In: 22nd International Conference on Engineering Mechanics 2016; May 2016; Svratka, CR, ITAM AS Prague; pp. 309–312. ISSN 1805–8248

[20] Králik J. Probabilistic nonlinear analysis of NPP hermetic cover failure, due to extreme pressure and temperature. In: Transactions of the VSB, Civil Engineering Series, No. 2, June 2015, Vol. 15, TU Ostrava; 2015. p. 14. DE GRUYTER OPEN ISSN 1804–4824.

[21] HANDBOOK 3. Implementation of Eurocodes, Action Effects for Buildings, Development of Skills Facilitating Implementation of Eurocodes. Leonardo Da Vinci Pilot Project CZ/02/B/F/PP-134007; Aachen, 10.2005

[22] HANDBOOK 5. Implementation of Eurocodes Reliability Backgrounds. Design of Buildings for the Fire Situation. Development of Skills Facilitating Implementation of Eurocodes. Leonardo Da Vinci Pilot Project CZ/02/B/F/PP-134007; Aachen; 2005

[23] NUREG/CR-7031. A Compilation of Elevated Temperature Concrete Material Property Data and Information for Use in Assessments of Nuclear Power Plant Reinforced Concrete Structures; ORNL/TM-2009/175. US NRC; 2010

[24] ACI 349–90. Code Requirements for Nuclear Safety Related Concrete Structures. Detroit: American Concrete Institute; 1990

[25] CEB-FIP Model Code. Design Code. Thomas Telford; 1990. ISBN: 0727716964

[26] Kohnke P. ANSYS, Theory. Canonsburg: SAS IP Inc.; 2008. www.ansys.com

[27] Kolmar W. Beschreibung der Kraftuebertragung ueber Risse in nichtlinearen Finite-Element-Berechnungen von Stahlbeton-tragwerken [PhD thesis]. Darmstadt: T.H. Darmstadt; 1986

[28] Jerga J, Križma M. Assessment of Concrete Damage, Building Research Journal 54. 2006. pp. 79–100. DE GRUYTER, ISSN 1339-682X

[29] Haldar A, Mahadevan S. Probability, Reliability and Statistical Methods in Engineering Design. New York, NY: John Wiley & Sons; 2000. https://www.amazon.com/Probability-Reliability-Statistical-Methods-Engineering/dp/0471331198

[30] JCSS 2011. JCSS Probabilistic Model Code. Zurich: Joint Committee on Structural Safety. <www.jcss.byg.dtu.dk>

[31] Bažant ZP, Pang SD, Vořechovský M, Novák D. Energetic-statistical size effect simulated by SFEM with stratified sampling and crack band model. International Journal for Numerical Methods in Engineering (Wiley). 2007;**71**(11):1297–1320. [Rep. Dept. of Civil Eng., Northwestern University, Evanston, Illinois]

[32] Novák D, Bergmeister K, Pukl R, Červenka V. Structural assessment and reliability analysis for existing engineering structures, Theoretical background. Structure and Infrastructure Engineering. 2009;**9**(2):267–275

[33] Melcher J, Kala Z, Holický M, Fajkus M, Rozlívka L. Design characteristics of structural steels based on statistical analysis of metallurgical products. Journal of Constructional Steel Research. 2004;60(3–5):795–808

[34] Kala Z, Melcher J, Puklický L. Material and geometrical characteristics of structural steels based on statistical analysis of metallurgical products. Journal of Civil Engineering and Management. 2009;**15**(3):299–307

[35] Krejsa M, Králik J. Probabilistic computational methods in structural failure analysis. Journal of Multiscale Modelling. 2015;**6**(2)p. 5. Imperial College Press, DOI: 10.1142/S1756973715500067

[36] Suchardová P, Bernatík A, Sucharda O. Assessment of loss results by means of multi - Criteria analysis. In: Advances in Safety, Reliability and Risk Management, Proceedings of the European Safety and Reliability Conference, ESREL 2011; London: CRC Press-Taylor & Francis Group; 2011. pp. 1563–1570. ISBN: 978-0-415-68379-1

[37] Sýkora M, Holický M. Assessment of Uncertainties in Mechanical Models, Applied Mechanics and Materials. Vol. 378. Switzerland: Trans Tech Publications; 2013. pp 13–18. doi:10.4028/www.scientific.net/AMM.378.13

[38] Vejvoda S, Keršner Z, Novák D, Teplý B. Probabilistic safety assessment of the steam generator cover. In: Proceedings of the 17th International Conference on Structural Mechanics in Reactor Technology (SMiRT 17); 2003 August 17–22; Prague, CR; M04-4; 2003. p. 10. https://www.iasmirt.org/smirt/17/transactions

[39] Králik J. Probabilistic nonlinear analysis of the hermetic circular cover of main coolant pump failure due to extreme pressure and temperature. In: 22nd International Conference on Engineering Mechanics 2016, Vol. 821; May 2016; Svratka, CR; 2016. pp. 309–312. ISSN 1805–8248

[40] Králik J. Probabilistic nonlinear analysis of NPP hermetic cover failure due to extreme pressure and temperature. In: Transactions of the VSB, Civil Engineering Series, No. 2, June 2017, Vol. 15, No. 2; TU Ostrava; 2017. p. 14. DE GRUYTER OPEN, ISSN 1804-4824

Chapter 6

Analysis of the Spatial Separation Effects of Thorium/Uranium Fuels in Block-Type HTRs

Ming Ding and Jie Huang

Additional information is available at the end of the chapter

http://dx.doi.org/10.5772/intechopen.68671

Abstract

With the rapid development of nuclear energy, thorium has been gaining attention because of its abundant reserves and excellent physical properties. Compared with light-water reactors, block-type high temperature gas cooled reactors (HTRs) are a better choice for thorium-based fuel for higher burnup and harder neutron spectrum. When using thorium in block-type HTRs, the composition and spatial distribution of thorium/uranium fuels are two determined factors of nuclear performance. Four spatial separation levels of thorium/uranium fuels, no separation level, TRISO level, channel level, and block level, are defined for the block-type thorium-fueled HTRs. A two-step calculation scheme was used to obtain the neutronic performance, including the initial inventory of U-235, effective multiplication factor, and average conversion ratio. Based on these data, the fuel cycle cost of different spatial separation levels can be calculated by the levelized lifetime cost method as a function of thorium content. The fuel cycle cost changes with the same trend as the initial inventory of U-235 in the reactor cores because the latter determines 70% of the total cost. When the thorium content is constant, the initial inventory of U-235 decreases with the increase of the spatial separation level because spatial self-shielding effect is strengthened by the latter.

Keywords: spatial separation effects, Th/U MOX, seed-and-blanket concept, whole assembly seed-and-blanket concept, block-type HTRs, fuel cycle cost analysis

1. Introduction

During the mid-1950s to the mid-1970s, some different types of Th/U MOX fuels, for example (Th,U)O_2 or (Th,U)C_2, were tested and used as fuels in high-temperature gas-cooled reactors (HTRs), like AVR and THTR in Germany [1] and Fort St. Vrain in the USA [2]. Three main reasons inspired them to demonstrate Th-U fuel cycle at that time. First, compared with kwon uranium, thorium is three times more abundant in nature. Second, Th-232 has an attractive potential for breeding to fissile U-233 efficiently in thermal neutron spectrum. U-233 is also considered as the best compared with other two common fissile isotopes, U-235 and Pu-239, in epithermal or thermal spectrum from neutronic point of view because the number of fission neutrons per neutron absorbed is 10–20% higher than that of U-235 and Pu-239. Finally, uranium resources were believed to be insufficient to support the development of nuclear industries on a large scale at the early period of nuclear energy development.

Besides HTRs, thorium has been an interesting nuclear fuel for various reactor applications [3], such as molten-salt reactors (MSRs) and water-cooled reactors.Especially, combined with molten-salt fuel in MSRs, thorium was used as a necessary composition of the "standard" fuel aiming to converse and even breed Th-232 to U-233 from the 1970s to mid-1980s. In past 20 years, the potential of thorium has been extended to radioactive waste management and plutonium incineration [4]. In this application, Th-232 is considered as a better fertile isotope than U-238 because of a larger net destruction of plutonium, when weapon-grade or reactor-grade plutonium is burned in various nuclear reactors. Furthermore, thorium-fueled reactors generate less long-lived radioactive wastes than uranium-fueled ones.

Combined with the commercial purpose and potential application scale, light-water reactors (LWRs) are a natural choice from the reactor point of view in recent years because of a large amount of LWRs all over the world. The seed-and-blanket (S&B) concept has been reexamined for LWR application by the MIT group [5, 6]. The concepts originate from S&B configuration and were developed for the advanced water breeder application (AWBA) program and tested in the light water breeder reactor (LWBR) at Shippingport from 1977 to 1982 [7]. Further work on thorium-fueled LWRs has been pursued by Radkowsky and Galperin [8]. The MIT group proposed the micro-heterogeneous fuel assemblies and the whole assembly seed-and-blanket (WASB) concept. Moreover, compared with the past research, the recent MIT work is all based on low-enriched uranium for proliferation resistance.

Because of the burnup limit of LWRs (usually 50 GWd/tHM), some research showed that the potential of thorium is limited in LWRs, and HTRs are a better choice for thorium-based fuel for higher burnup and harder neutron spectrum. Recently, the concept of S&B fuel blockshas been introducedto the U-battery [9], a small long-life HTR, and a commercial-level GT-MHR [10], as well as advanced high-temperature reactors (AHTRs) [11] with low-pressure liquid-salt (Flibe) as coolant, which enables the design of a high-power (e.g., 2400–4000 MWth), high-temperature (850–950°C) reactor with fully passive safety capability and the economic production of electricity orhydrogen.

The past studies have shown the distribution of thorium and uranium fuels in space is a very important factor for the performance of thorium-fueled reactors and have proposed some interesting concepts of S&B and WASB. Actually, the different concepts represent different separation levels of thorium/uranium fuels in space. The chapter tries to systematically analyze the spatial separation effects of thorium/uranium fuels on the nuclear performance of block-type HTRs and to quantitatively evaluate the difference by fuel cycle cost. The second section will describe the reactor, fuel block, and the calculation models, including the transport calculation model with burnup and fuel cycle cost model. The third section will present the calculation results of four different spatial separation levels and discuss the difference among them. The final section will conclude the chapter.

2. Block-type HTRs and calculation methodologies

2.1. Block-type HTRs and spatial separation levels

2.1.1. Block-type HTRs

Modular block-type HTRs [12] are one kind of inherently safe reactors, which passively remove decay heat by natural convection, conduction, and radiation from the reactor core and keep the fuel intact. As mentioned in Section 1, the block-type HTRs are a better choice for thorium-based fuel for higher burnup and harder neutron spectrum. Moreover, an advantage of block-type HTRs over pebble-bed HTRs is that the former provides multi-level and fixed spatial distribution of thorium/uranium fuels in the reactor cores.The reactor core of the block-type HTRs investigated, as shown in **Figure 1(a)**, is comprised of an annular fuel zone, inner reflectors, and outer reflectors. The annular fuel zone is comprised of 1980 fuel blocks, as shown in **Figure 1(b)**. The fuel blocks are stacked firmly against each other in columns that form an annulus between an inner and an outer reflector, both of which consist of rings of unfueled graphite blocks. The annular core configuration ensures higher thermal power and inherent safety under most accidents and transient conditions. Each fuel block includes 210 fuel channels, 108 coolant channels, and 6 fixed burnable poison channels which are filled with

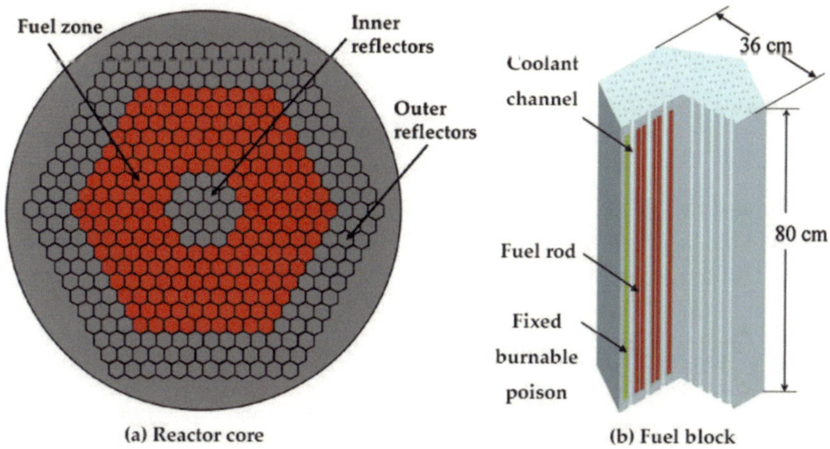

Figure 1. Thorium-fueled HTR and fuel block, (a) Reactor core and (b) Fuel block. Does the query aim to use second style for the book?

graphite in the flowing calculations.The cylindrical fuel compacts are stacked inside channels drilled into hexagonal graphite blocks. Other geometrical parameters ofthe fuel blocks are listed in **Table 1**.

The most prominent feature of the block-type HTRs is use of tristructural-isotropic (TRISO) particles, which contain five layers from inside to outside: fuel kernel, porous carbon buffer, inner pyrolytic carbon (IPyC), silicon carbide (SiC), and outer pyrolytic carbon (OPyC). The porous carbon buffer layer is provided to protect the dense PyC from fission recoil damage and to provide void volume to limit the fission gas pressure. The dense PyC has good irradiation stability and capability to remain intact to perform fission production retention under severe exposure conditions. Sandwiched between the two PyC layers, the SiC layer provides metallic fission production retention and mechanical strength. TRISO fuel particles are blended and bonded together with a graphite matrix to form fuel compacts. **Figure 2** illustrates how the TRISO particles are packaged into the annular core.

2.1.2. Definition of multiscale spatial separation

When using thorium in HTRs, two important factors influence nuclear performance. One is the thorium content (the mass ratio of Th-232 to all heavy metal isotopes), meaning the mixed proportion of thorium and uranium. The other is the spatial separation level of thorium and uranium, meaning how thorium and uranium are to be mixed. According to the level of spatial separation, four levels are in a block-type HTR: (1) No separation (Th/U MOX level): with Th/U MOX fuel, the thorium and uranium are mixed in each fuel kernel as a form of $(Th,U)O_2$, as shown in **Figure 2(a)**. Thorium and uranium do not separate in the macrolevel, (2) TRISO-level separation (SBT level): UO_2 and ThO_2 are made into different TRISO fuel particles (UO_2 TRISO and ThO_2 TRISO) separately, but the two kinds of TRISO fuel particles are mixed into the same fuel compact, as shown in **Figure 2(b)**, (3) Channel-level separation (SBU level): each fuel channel has only one kind of fuel compacts (UO_2 fuel compacts or ThO_2 fuel compacts), but a fuel block has both UO_2 fuel channels and ThO_2 fuel channels, as shown in **Figure 2(c)**, and (4) Block-level separation (WASB level): each fuel block only has a kind of fuel (UO_2 or ThO_2), but the core has both UO_2 fuel blocks and ThO_2 fuel blocks, as shown in **Figure 2(d)**.

2.2. Calculation and evaluation methods

2.2.1. Neutronic calculation method

To only analyze the influence of spatial separation, the other parameters are kept the same including the fuel block geometry, as listed in **Table 1**, and the fuel shuffling scheme, as shown in **Figure 3**. A detailed full-core 3D transport calculation in one step will require significant

Parameters	Width across flat [cm]	Height of block [cm]	Diameter/ number of fuel channels [cm/−]	Diameter/number of coolant channels [cm/−]	Diameter of kernels [μm]	Thickness of TRISO layers [μm]	Density of TRISO layers [g/cm^3]
Values	36	79.3	1.27/210	1.588/102, 1.27/6	500	100/35/35/40	1.05/1.9/3.2/ 1.9

Table 1. Basic geometrical parameters of fuel blocks in the HTR.

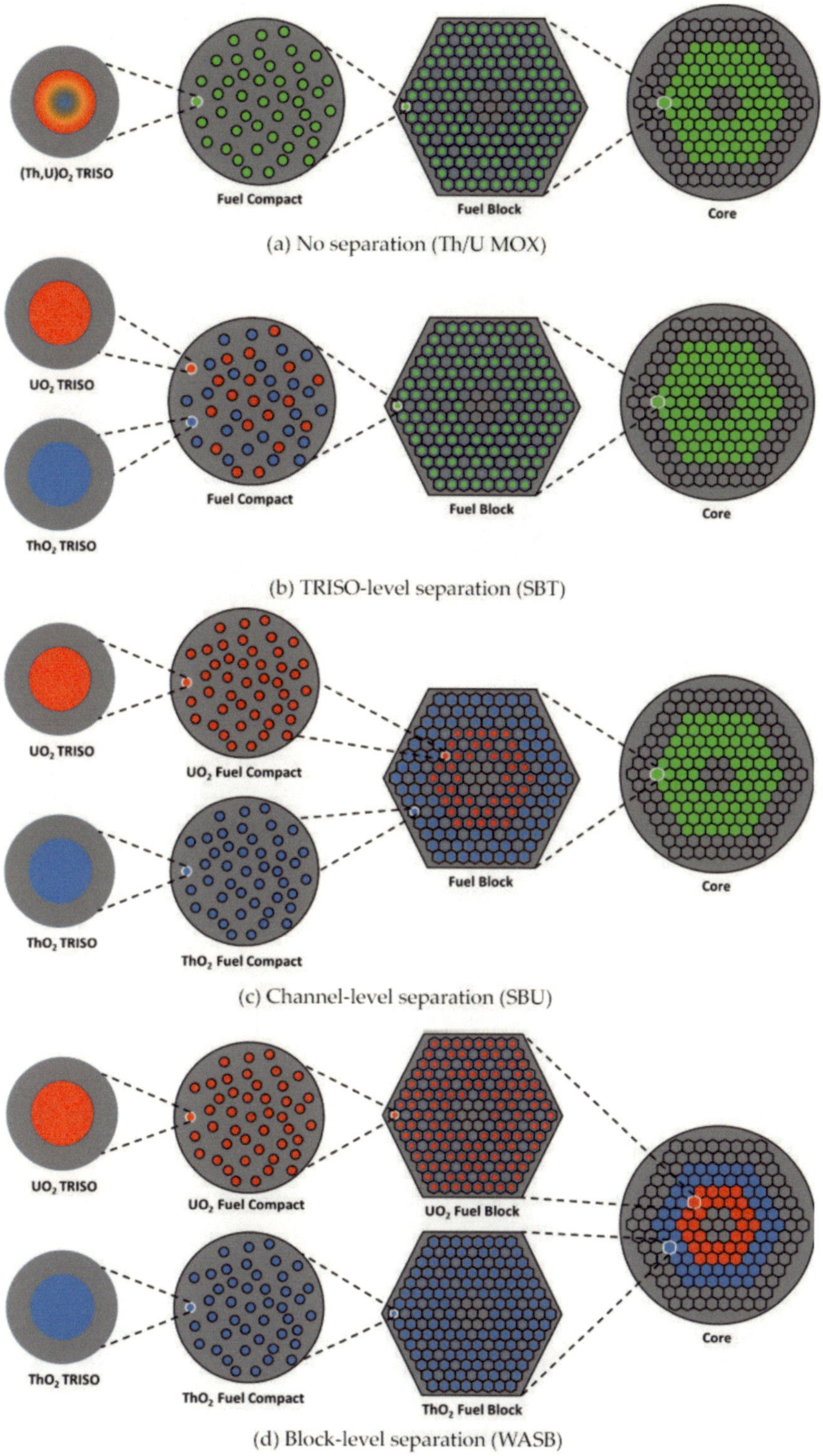

Figure 2. Four levels of spatial separation of thorium/uranium in block-type HTRs.

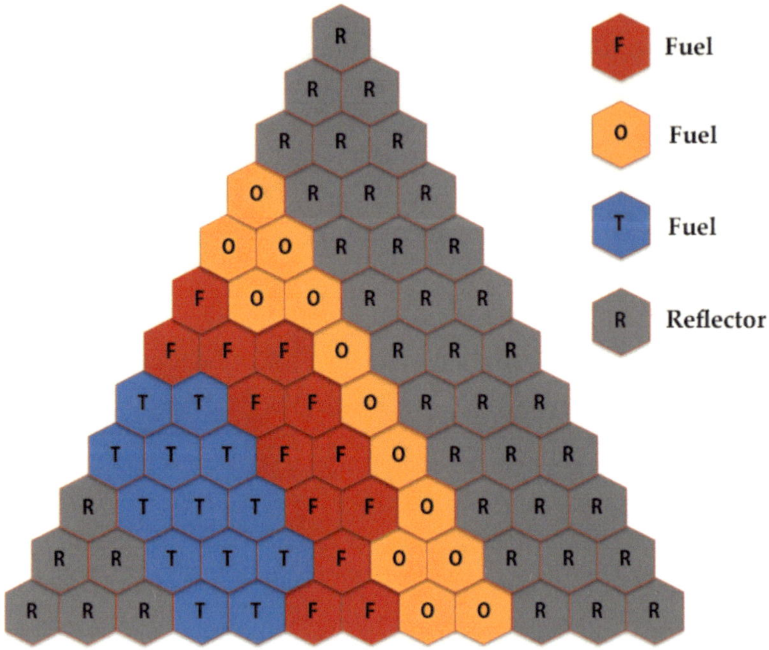

Figure 3. Reactor core calculation model for 1/6th core.

memory and a central processing unit (CPU) time as utilities in producing calculations or even in laboratories for design purposes. Therefore, a two-step calculation scheme is typical: (1) a detailed calculation at the assembly level with reflective boundary conditions, which gives homogenized cross-sections for the assemblies, condensed to a certain number of groups (lattice calculation step) and (2) a second calculation at the core level with homogenized properties in each assembly and usually small number of groups (full-core calculation step).

In this chapter, the traditional two-step calculation scheme is constructed based on the DRAGON V4 code system [13], as shown in **Figure 4**. An attractive feature of the DRAGON code is its ability to treat particle fuel in a graphite matrix in a full-assembly calculation. It provides the possibility to define a stochastic mixture of spherical micro-structures that can be distributed inside composite mixtures of the current macro-geometry using the Hebert double-heterogeneity model [14] or Sanchez-Pogroming double-heterogeneity model [15].

In step (a) in **Figure 4**, a 2D fuel block is modeled including the structure of the TRISO particle. The method of characteristics (MOC) and 295-group cross-section library (http://www.polymtl.ca/merlin/libraries.htm) are used to solve the transport equation. Then, the energy groups of the cross-sections are condensed to 26 groups during the homogenization of the fuel block. In step (b), a 1D annular core is modeled, comprising a homogenized fuel zone and two reflector zones. The homogenized cross-sections of the fuel block generated in step (a) are used for the fuel zone in this step, and the cross-sections of the graphite generated in step (a) are used for the reflectors of the 1D reactor model. The 1D transport calculations are executed by the discrete ordinate (S_N) method. Only the reflectors are homogenized to generate the

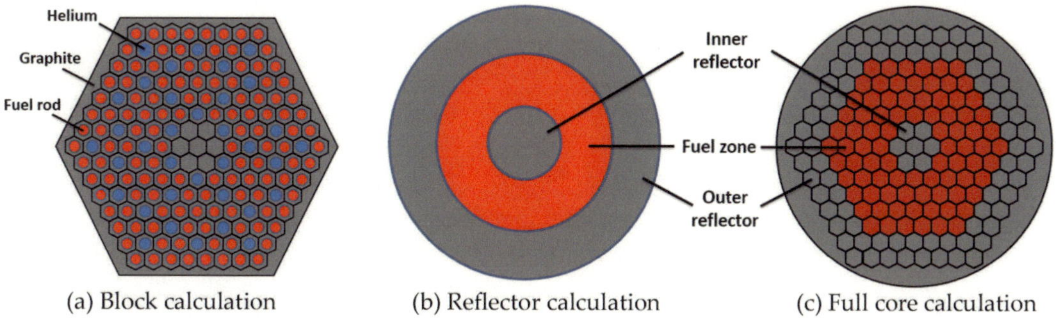

Figure 4. A schematic diagram of the two-step calculation scheme.

cross-sections of the reflector for the following full-core calculations. In step (c), the 2D full-core calculations are executed by the interface current (IC) method. The cross-sections of fuel blocks in this step come from the step (a), while the cross-sections of reflectors come from step (b).

2.2.2. *Fuel cost analysis model*

An evaluation criterion must be in the comparison of different spatial separationlevels. Usually, the fuel cycle cost is one of the criteria. To make a fair comparison, the adoption of a standardized methodology for fuel cycle cost calculations is prerequisite. Therefore, the levelized lifetime cost methodology is used in this study. The levelized lifetime cost methodology was developed by the Organization for Economic Cooperation and Development (OECD) [16], which uses an internationally accepted investment appraisal methodology to analyze the fuel cycle cost.

The cash flow for fuel cycle material and services commences before the reactor starts to generate electricity and continues well after the reactor ceases operation. The exact timing of payments for natural uranium purchase, conversion, enrichment, fabrication, spent fuel storage, and disposal depends on the associated lead and lag times for the fuel cycle components. To calculate the overall fuel cycle cost, the magnitude of each component cost and the appropriate time that it occurs must be identified. Fuel quantities are obtained from reactor neutronic calculations as described in Section 2.2.1. These quantities of materials and services are adjusted to allow for process losses in the various component stages of the nuclear fuel cycle and then multiplied by the unit costs to obtain the component costs. Finally, the total fuel cycle cost is gotten by discounting these component costs to their present values in a specified base year (usually the commission date of reactor), as shown in **Figure 5**. The levelized lifetime cost methodology provides costs per unit of electricity generated which are the ratios of lifetime expenses to total expected output, expressed in terms of the present value equivalent. This method derives economic merits by comparing their respective average levelized lifetime costs.

The reactor data and fuel cycle cost data for the HTR are shown in **Tables 2** and **3**, respectively. **Table 2** gives the information about the reactor including the power, lifetime, cycle length, and back-end options. **Table 3** gives the unit price and lead or lag time for natural uranium

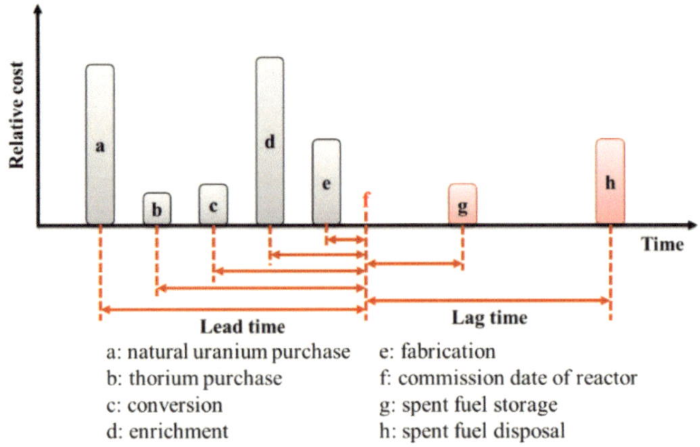

Figure 5. Time flow of nuclear fuel cycle cost with direct disposal option.

Items	Values
Thermal output	600 MWth
Electric output	240 MWe
Plant lifetime	60 years
Refueling cycle length	2 years
Load factor	90%
Back-end options	Direct disposal
Discount rate	8%

Table 2. Reactor operation data for the HTR.

Components	Prices	Lead or lag times
Natural uranium purchase	120.9 $/kg U	24 months
Thorium purchase	120.9 $/kg Th	24 months
Conversion	8.75 $/kg U	18 months
Enrichment	126.5 $/SWU	12 months
Fabrication	777 $/kg HM	6 months
Storage	230 $/kg HM	5 years
Disposal	610 $/kg HM	40 years

Table 3. Fuel cycle cost data for the HTR.

purchase, thorium purchase, conversion, and enrichment. The price of natural uranium, conversion, and enrichment is the average price of spot market from 2010 to 2015 (http://www.uxc.com/), respectively: 120.9 $/kg U, 8.75 $/kg U, and 126.5 $/SWU. The fabrication cost of 777 $/kg HM for HTR fuel blocks comes from the Oak Ridge National Laboratory (ORNL)'s report [17]. Others are the same as the OECD's report [16]. Although different fuel separation levels would likely affect unit price related to fabrication and, in particular, storage, and disposal, their fraction in total fuel cycle cost is small, as mentioned in Section 3.4. Moreover, the sensitivity analysis shows that a ±50% variation of the price of fabrication, storage, or disposal will lead to a no-more-than ±5% variation of fuel cycle cost [18]. Thus, even if different fuel separation levels have a different price of fabrication, storage, and disposal, the analysis in this chapter is still valid.

3. Results and discussion

For real HTRs with multi-batch refueling and multi-block, both the spatial separation levels defined in Section 2 and refueling patterns influence the performance of fuel in the reactor core, so the latter is excluded by ideally adopting one-batch fixed-patternrefueling mode in order to focus on the spatial separation effect of the thorium/uranium fuel. Based on the one-batch fixed-pattern refueling mode, as shown in **Figure 3**, thorium and uranium in the different spatial separation levels with the same mass are loaded into the geometrically same reactor core and are discharged when the k_{eff} reaches the same specified value. The core performance difference is due to the spatial separation levels.

3.1. Configuration effects in one spatial separation level

In order to compare four different spatial separation levels in the core scale, the configurations or patterns of the thorium/uranium fuels in the channel-level separation and block-level separation are simply analyzedin this section because the former meets different configurationsor patterns in block scale and the latter does in core scale, as shown in **Figure 2**, and the configurations influence the performance of the reactor core. For the other two separation levels, that is no separation level and TRISO-level separation, the same the fuel blocks exist in the whole reactor core.

3.1.1. Spatial configuration effects in the channel-level separation

For the channel-level separation, the spatial configurations of thorium/uranium compacts in the fuel block are different even for the same thorium content because uranium and thorium compacts are located in different channels, that is, there are 2^{210} configurations theoretically. It is almost impossible to investigate all of them, but it is not necessary to do it because the performance is possibly similar for some similar configurations, and the comparison in the reactor core scale is of the main concern. Five typical configurations (SBU 1#–SBU 5#) are chosen and investigated for 46% of thorium content, as shown in **Figure 6**. The uranium fuel compacts are located in the central, middle, and outer regions of the fuel block, respectively, for the SBU 1#–SBU 3#, and are evenly located in the fuel block for the SBU 4# and SBU 5#. The uranium compacts are relatively concentrated in the former group of configurations and are relatively dispersed in the latter group of configurations. For other thorium content, the spatial separation of uranium/thorium compacts are located in the similar mode.

Figures 7 and **8** show the k_{eff} of the reactor core for 46 and 91% thorium contents, which are calculated based on the model described in Section 2.2. When the thorium content increases, the number of thorium compacts will increase and the number of uranium compacts will decrease. In order to maintain the same mass of U-235 in the fuel block or in the reactor core,

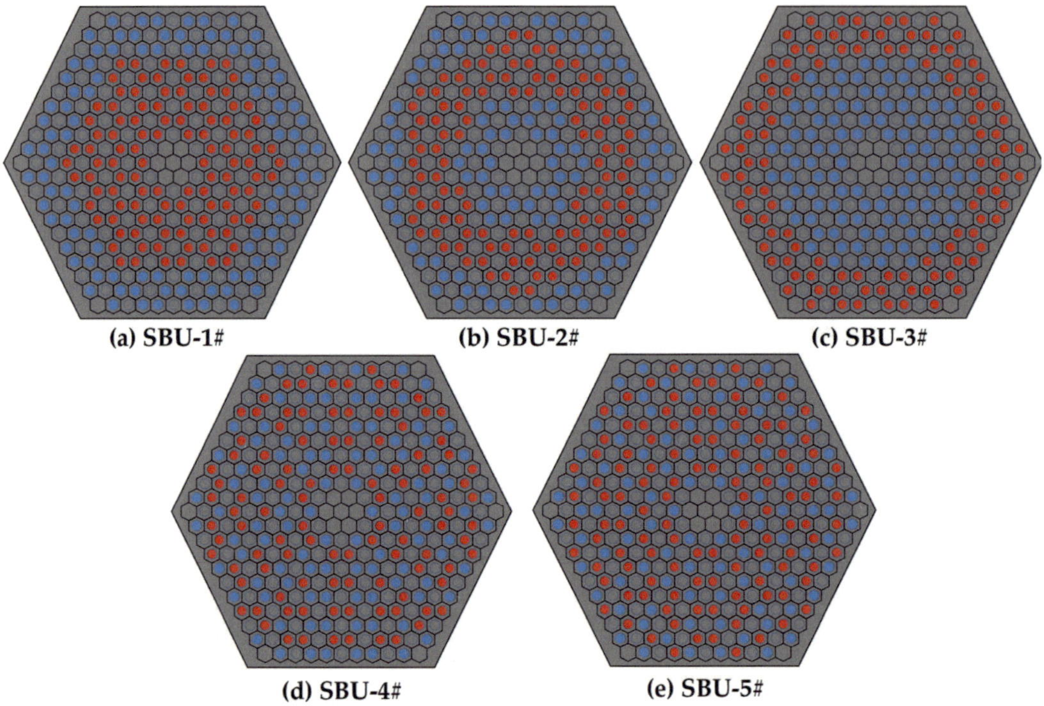

Figure 6. Five typical configurations of uranium/thorium fuels for 46% thorium content (red: uranium compacts, blue: Thorium compacts).

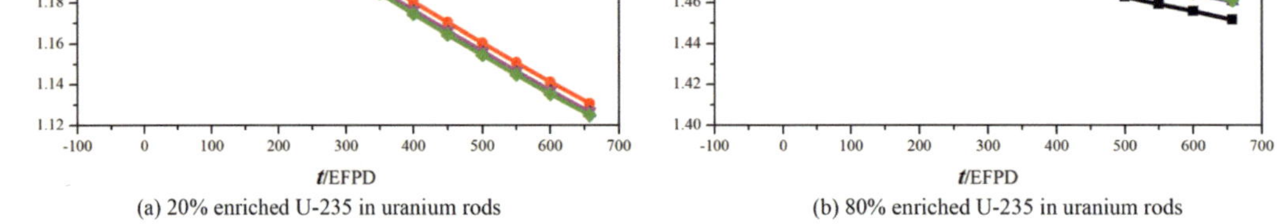

Figure 7. k_{eff} of five configurations for 46% thorium content.

the enrichment of U-235 has to be increased. For different thorium contents, when the enrichment of U-235 is less than 20%, the five different spatial configurations of thorium/uranium fuels achieve the similar k_{eff}. However, when the enrichment is higher than 20%, for example 80%, the k_{eff} of the five configurations is different obviously. Moreover, the k_{eff} of the SBU 4# and SBU 5# is always higher than the other three configurations, and the influence of the spatial configurations becomes important. In the channel-level separation, the dispersed uranium compacts are advantageous to transmute Th-232 to U-233 when the thorium content and, thus, the enrichment of U-235 are high. Based on the calculations and analysis, the suggested spatial configurations of thorium/uranium compacts in the fuel block are shown in **Figure 9** for the channel-level separation. The uranium compacts are concentrated in the

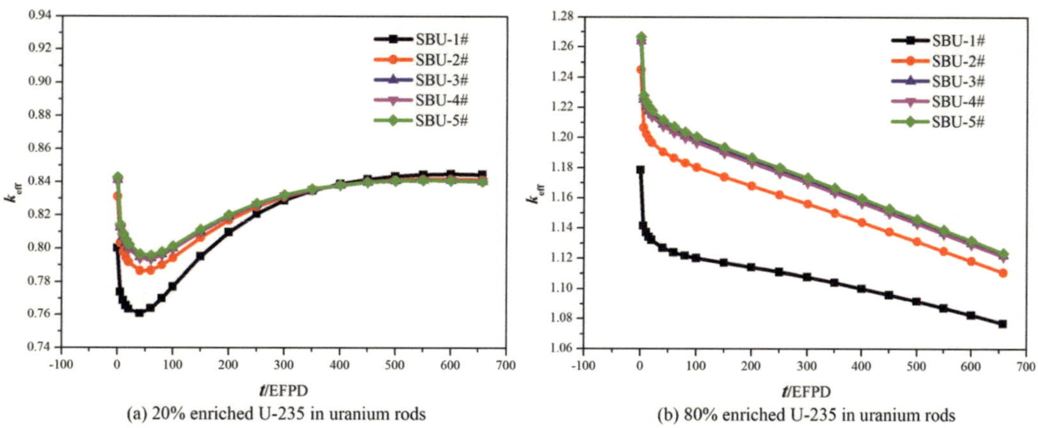

Figure 8. k_{eff} of five configurations for 91% thorium content.

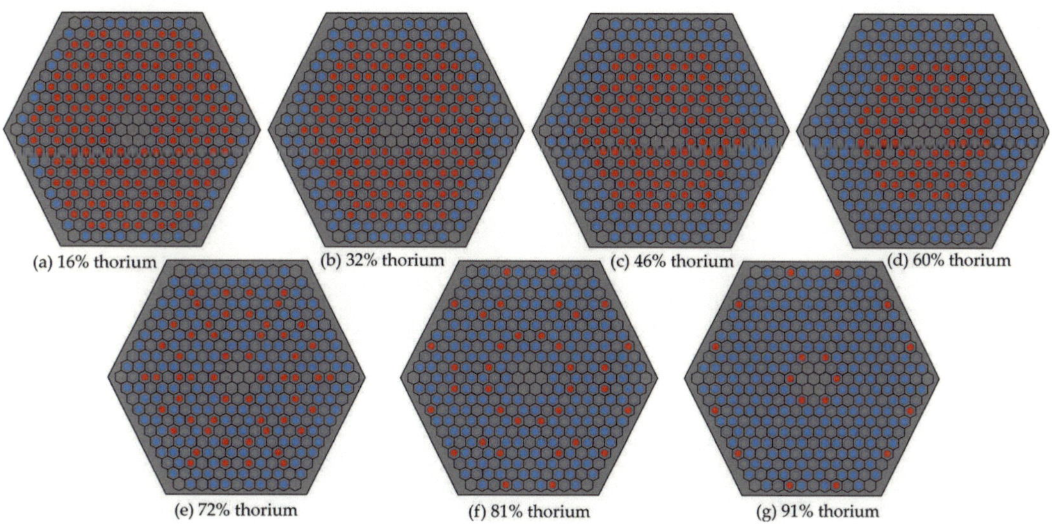

Figure 9. Configurations of thorium/uranium compacts for channel-level separation (red: uranium compacts, blue: thorium compacts).

central region of the fuel block for low thorium content and are dispersed for high thorium content.

3.1.2. Spatial configuration effects in block-level separation

For the block-level separation, the spatial configurations of thorium/uranium fuel blocks in the reactor coremeet the same problem as the channel-level separation. The uranium and thorium fuel blocks located in different positions in the reactor core lead to different configurations, that is, there are 2^{198} configurationsof the fuel blocks theoretically. Based on the same analysis method as presented in Section 3.1.1, every five typical configurations (WASB 1#–WASB 5#) were chosen and investigated for different thorium content. The results show that the dispersed thorium fuel blocks are advantageous to transmute Th-232 to U-233 in the block-level separation. Based on the results, the suggested spatial configurations of thorium/uranium fuel blocks in the reactor core are shown in **Figure 10** for the block-level separation, which is chosen to be compared with other three separation levels.

3.2. Effective enrichment and initial inventory of U-235

As mentioned in Section 3.1, the highly-enriched U-235 possibly has to be used when the thorium content increases. Compared with thorium/uranium-fueled reactor, the uranium-fueled reactor only contains one fissile isotope, that is, U-235, and one fertile isotope, that is,

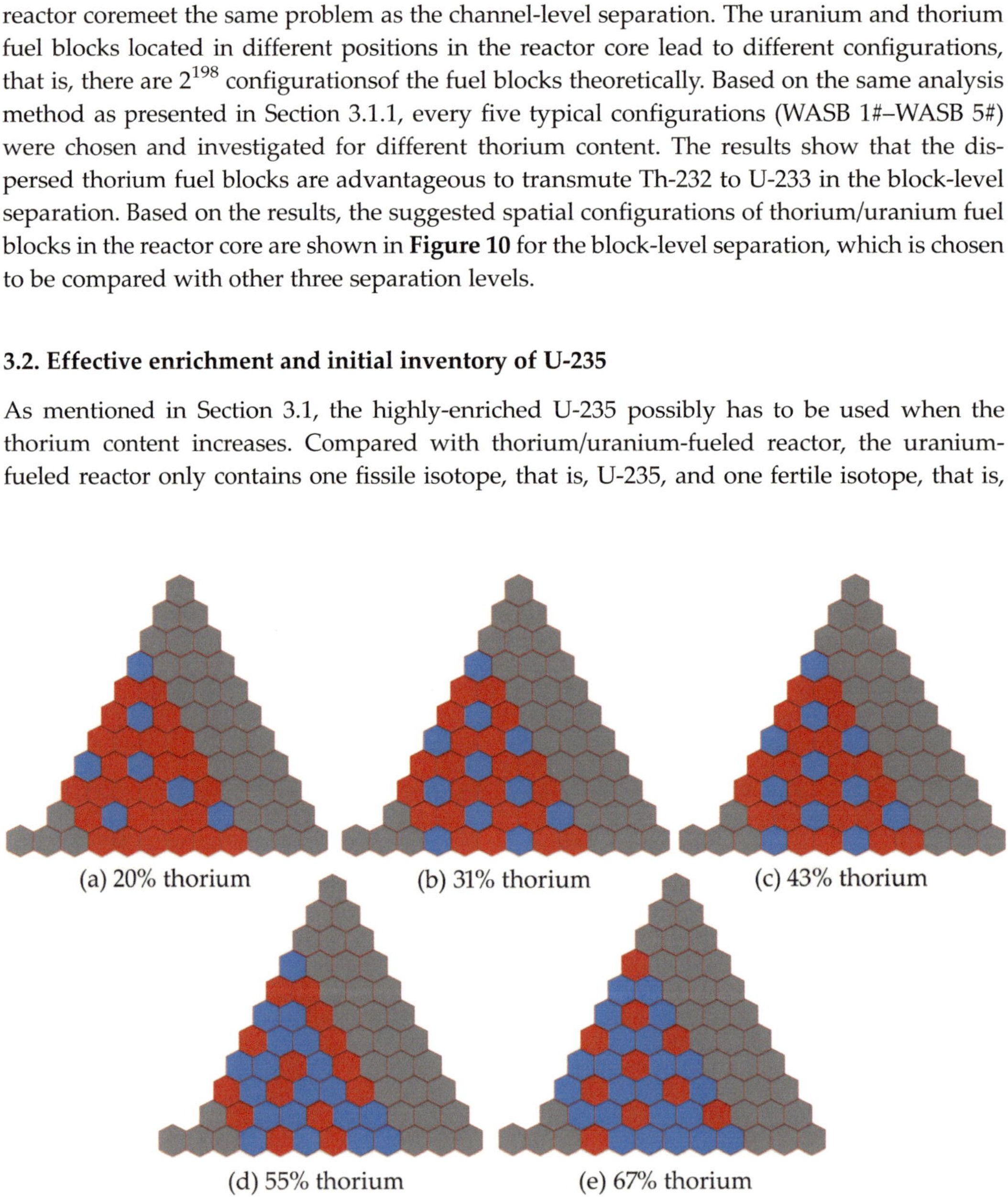

Figure 10. Configurations of thorium/uranium fuel blocks for block-level separation (red: uranium fuel blocks, blue: thorium fuel blocks).

U-238, in the fuel. The mass fraction of fissile isotopes, that is, U-235, is the so-called enrichment of U-235, as defined in Eq. (1).,

$$\varepsilon = \frac{m_{\mathrm{U5}}}{m_{\mathrm{U5}} + m_{\mathrm{U8}}} \tag{1}$$

When another fertile isotope, that is, Th-232, appears in the reactor fuel, the mass fraction of fissile isotopes, as defined in Eq. (2)

$$\varepsilon_{eff} = \frac{m_{\mathrm{U5}}}{m_{\mathrm{U5}} + m_{\mathrm{U8}} + m_{\mathrm{Th2}}} \tag{2}$$

is different from the traditional enrichment of U-235, which is called effective enrichment of U-235[19]. If the concept of the effective enrichment of U-235 is adopted, the nominal mass fraction of fissile isotopes is far less than 20% for the thorium-fueled HTRs as shown in **Figure 11**, which is the limit enrichment of U-235 for low-enriched fuel. Since the enrichment of U-235 is no longer equal to the effective enrichment of U-235 in a thorium-loaded reactor, and the physical meaning of the latter is clearer, the latter is analyzed instead of the former.

As shown in **Figure 11**, the required initial effective enrichment of U-235 obviously decreases with the increase of separation level from Th/U MOX to SBU and WASB for the same 2-year

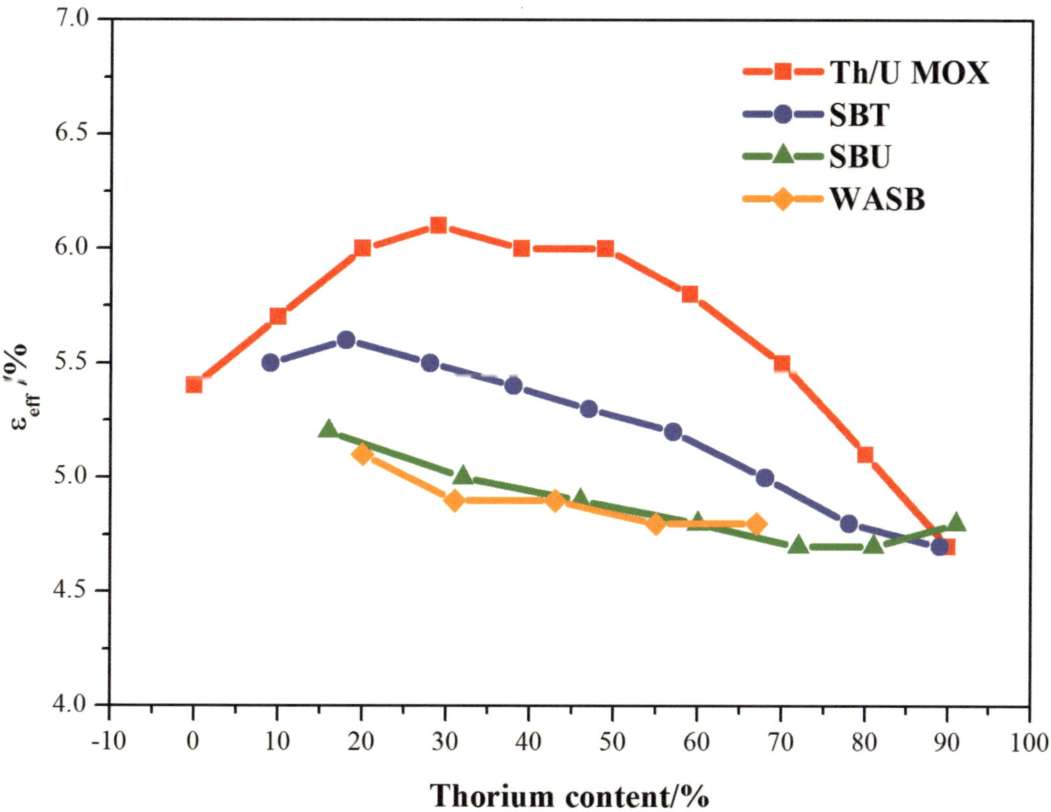

Figure 11. Initial effective enrichment of U-235 as a function of thorium content for different separation levels.

refueling cycle length, when the thorium content changes from 10 to 80%. When the thorium content is less than 10% or larger than 80%, the influence of spatial separation level on the initial ε_{eff} is weak because the fuel is nearly uranium or thorium and thus there is no obvious difference. Moreover, the initial ε_{eff} of SBU is nearly the same as that of theWASB when the thorium content is in the range of 20–70%. For the Th/U MOX fuel, the initial ε_{eff} increases when the thorium content increases from 0 to 30% and decreases with the further increase of thorium. For the SBT, SBU, and WASB, the initial ε_{eff} decreases with the increase of the thorium content.

Although the total initial inventory of heavy metal slightly decreases because of the density difference between ThO_2 (9.4 g/cm^3) and UO_2 (10.4 g/cm^3), as shown in **Figure 12**, the required initial inventory of U-235,as shown in **Figure 13**, nearly has the same trend as the initial effective enrichment of U-235, according to Eq. (2). Furthermore, the initial inventory of the enriched uranium also decreases with the increase of the thorium content, as shown in **Figure 12**, because more and more U-238 is replaced by Th-232 in the reactor core. The change of the initial inventory of U-235 or the initial effective enrichment of U-235 is a result of the difference of nuclear performance (e.g., initial effective multiplication factor and average conversion ratio) caused by the spatial separation levels, as furtherly discussed in Section 3.3.

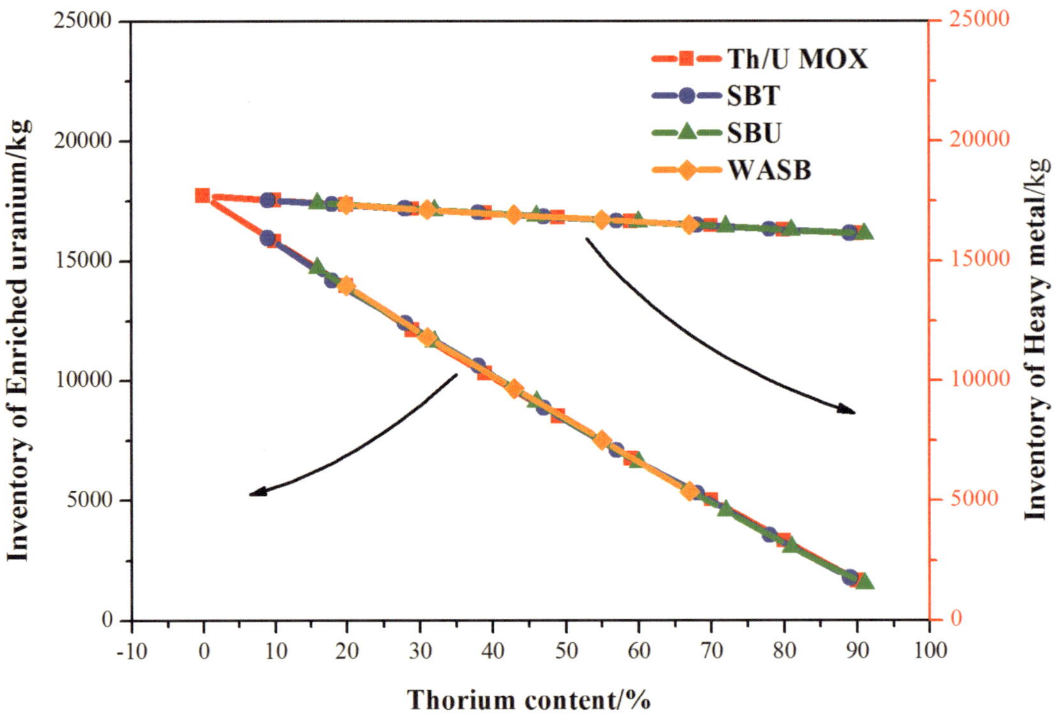

Figure 12. Initial inventory of heavy metal as a function of thorium content for different separation levels.

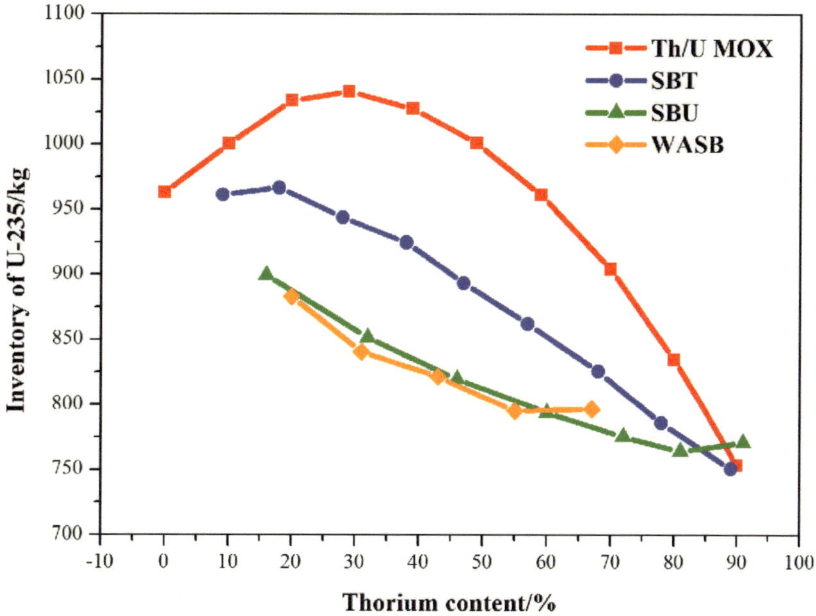

Figure 13. Initial inventory of U-235 as a function of thorium content for different separation levels.

3.3. Initial effective multiplication factor

3.3.1. Initial effective multiplication factor and average conversion ratio

Figure 14 presents the initial effective multiplication factors (k_{eff}) and average conversion ratios (ACRs) of four spatial separation levels as a function of thorium content. For each spatial separation level, the k_{eff} nearly decreases with the increase of thorium content. For the same thorium content, the initial k_{eff} usually increases when the spatial separation level increases

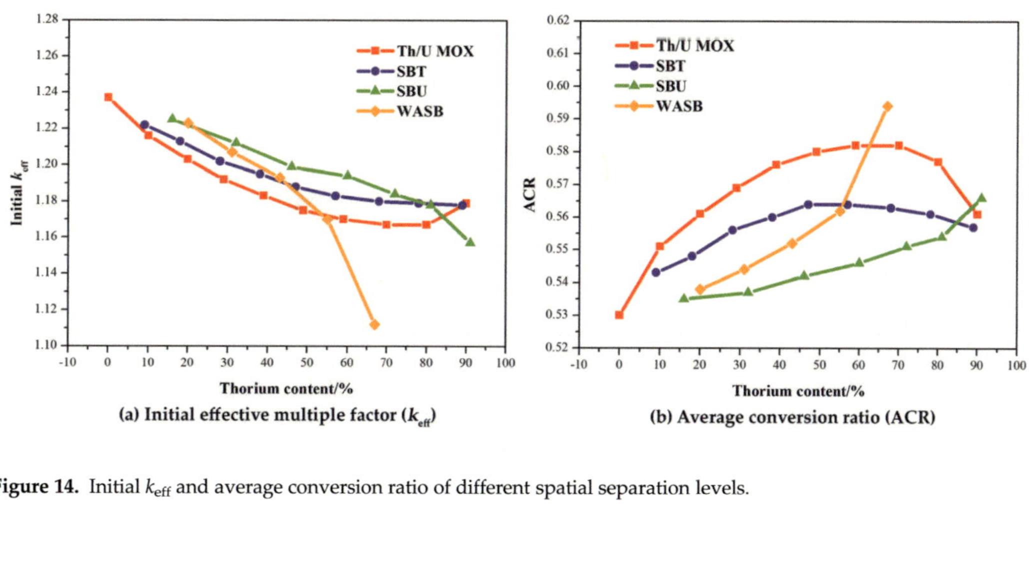

Figure 14. Initial k_{eff} and average conversion ratio of different spatial separation levels.

from Th/U MOX to SBU. When the thorium content is larger than 80%, the difference among them becomes small. The more interesting rule is that the trend of ACR is always opposite to the initial k_{eff}, that is, when the ACR is higher, initial k_{eff} is smaller, even for the WASB level.

Figure 15 presents the k_{eff} as a function of operation time for two different ACRs in one refueling cycle. In these calculations, the refueling cycle length is 657 effective full power days (EFPDs), which means the coordinate of end of cycle (EOC) is fixed, as the point B (657,1.005) in **Figure 15**. On the other hand, the k_{eff} is a nearly linear function of burnup for the thorium-fueled HTR as shown in **Figure 15**. Therefore, if the average conversion ratio of the refueling cycle is smaller, the reactivity drop is larger and thus the initial k_{eff} must be higher to guarantee a critical reactor at EOC.

If **Figure 14** is compared with **Figure 11** or **Figure 13**, it is interesting to find that when the spatial separation level of thorium/uranium fuels changes from SBU to Th/U MOX, the required initial inventory of U-235 is the most for the Th/U MOX in order to achieve the same operation time, but the initial k_{eff} is smallest and the main benefit is the most amount of Th-232 transmuted into U-233. Once the spatial separation of thorium/uranium fuels changes from SBU to WASB, the required initial inventory of U-235 decreases. Moreover, the ACR increases

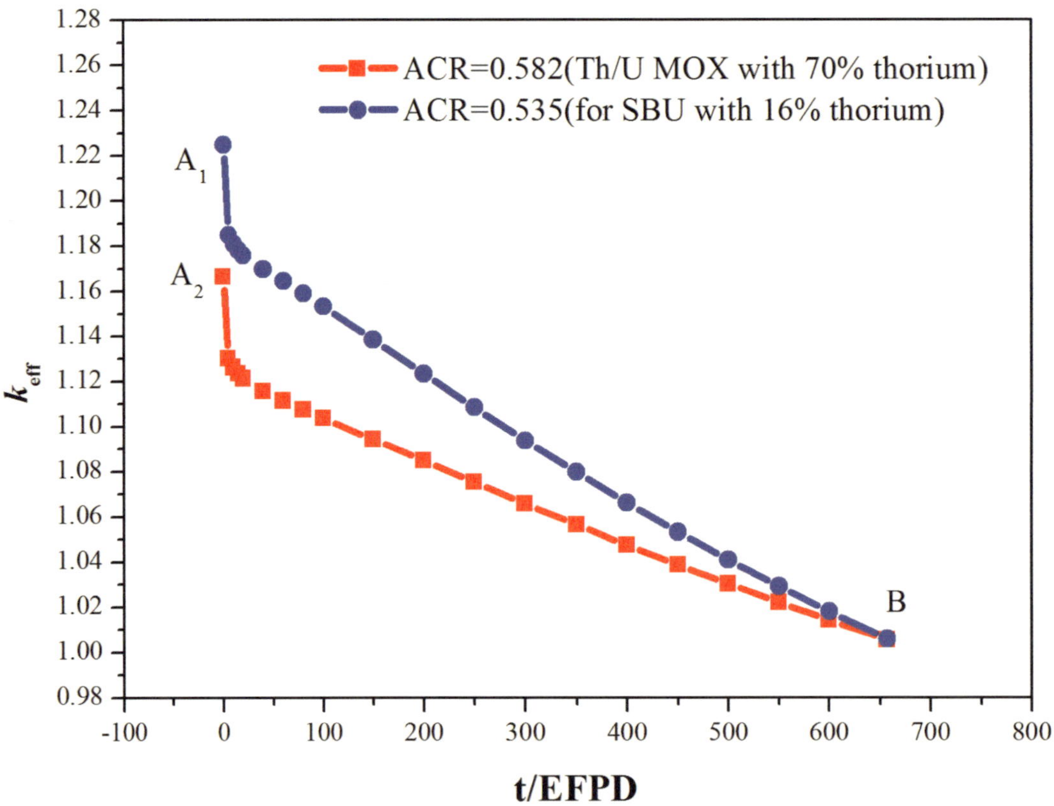

Figure 15. k_{eff} as a function of EFPD for different average conversion ratios.

with increase of thorium content, the initial k_{eff} decreases, and thus the reactivity drop or swingdecreases in the refueling length.

3.3.2. *Further discussion of initial effective multiplication factor*

In order to analyze the influence of spatial separation levels and thorium content on the initial k_{eff} and thus the initial inventory of U-235, four control groups of typical reactor cores including spatial separation levels and thorium content are calculated and discussed. The group 1 is to compare the Th/U MOX with SBU with 50% thorium content, in order to explain the reason of the decrease of the initial inventory of U-235. The group 2 is to compare the SBU and WASB with 50% thorium content, in order to explain the difference between them. The groups 3 and 4 are to discuss the Th/U MOX with 0, 30, and 80% thorium content, respectively, in order to explain the influence of the thorium content.

Table 4 presents the five factors and initial k_{eff} of the reactor cores involved in the four control groups, which are calculated according to the validated method [10]. The five-factor formula [20] can be written as

$$k_{eff} = \eta f p \varepsilon P_{NL} \tag{3}$$

where η, f, p, ε, P_{NL} are, respectively, the reproduction factor, fuel utilization factor, resonance escape probability, fast fission factor, and nonleakage probability. However, in the core physics calculation described in Section 2.2.1, the homogenization of fuel block will mix the fuel, moderator, and coolant into a mixture, making the η and f inseparable. Thus, the production of η and f is regarded as the reproduction factor of the core. More information about the five-factor formula can be found in Ref. [10].

To quantitatively describe the contribution of each factor to the variation of the initial effective multiplication factor, its contribution [21] is defined in terms of the components of five-factor formula

$$\Delta r = \left(\frac{1}{\frac{r_1}{r} \cdot k} - \frac{1}{\frac{r_2}{r} \cdot k} \right) \cdot 10^5, \qquad r - \eta f, \varepsilon, p, P_{NL} \tag{4}$$

Reactor core	ηf	ε	p	P_{NL}	k_{eff}
Th/U MOX-0%Th	1.7257	1.0804	0.6994	0.9550	1.24531
Th/U MOX-30%Th	1.6891	1.0861	0.6852	0.9551	1.20058
Th/U MOX-80%Th	1.5540	1.0701	0.7379	0.9570	1.17436
Th/U MOX-50%Th	1.6491	1.0830	0.6935	0.9556	1.18358
SBU-50%Th	1.5690	1.0687	0.7508	0.9567	1.20442
WASB-50%Th	1.5182	1.0734	0.7567	0.9615	1.18567

Table 4. Five factors and initial k_{eff} of different reactor cores involved in four control groups.

where

$$\bar{r} = 0.5(r_1 + r_2), \quad \bar{k} = \overline{\eta f}\,\bar{\varepsilon}\,\bar{p}\overline{P_{\text{NL}}} \tag{5}$$

The subscripts 1 and 2 represent two different cases, respectively. **Table 5** presents the contribution of each factor for the four control groups.

For the group 1, when the thorium content is 50% and spatial separation level changes from Th/U MOX level to SBU level, the nonleakage probability increases by 96 pcm and contributes +6% to the initial k_{eff}, because the spatial separation level strengthens thorium absorption, and the microscopic absorption cross section of Th-232 (7.4 barns) is three times of U-238 (2.7 barns). Based on the same reason, more thermal neutrons are absorbed by Th-232 but cannot induce fission reactions, which leads to the reproduction factor of core (ηf) to decrease by 4167 pcm and contributes −285% to the initial k_{eff}. The fast fission cross-section of Th-232 (0.01 barns) is smaller than that of U-238 (0.04 barns), which causes the fast fission factor (ε) to decrease by 1112 pcm and contribute −76%. However, because the thorium fuel is lumped in the thorium compacts for the SBU-50%Th and a smaller resonance integral (RI) of Th-232 (85) compared to U-238's (275), the resonance escape probability (p) of the SBU-50%Th increases by 6649 pcm and contributes +455%. As a result of all effects, especially the contribution of the increase of p, the initial k_{eff} of the SBU-50%Th is 1460 pcm larger than that of the Th/U MOX-50%Th. The spatial self-shielding effect is strengthened by Th-232 and spatial separation levels, and thus the increase of resonance escape probability leads to the decrease of 188 kg initial inventory of U-235.

For the group 2, when the spatial separation increases from SBU level to WASB level, although the resonance escape probability further increases by 655 pcm, the reproduction factor decreases by 2754 pcm because of the further lumping of the thorium fuel. As a result, the initial k_{eff} decreases by 1313 pcm causing a smaller reactivity swing in the refueling period. Moreover, the initial inventory of U-235 decreases 6 kg because of a larger ACR.

For the group 3, for a fixed spatial separation level, for example, Th/U MOX level, when the thorium content increases from 0 to 30%, more thermal neutrons generated by fission are absorbed by fertile Th-232 because of three-time microscopic absorption cross section of

Group number	Group	$\Delta\eta f$ [pcm]	$\Delta\varepsilon$ [pcm]	Δp [pcm]	ΔP_{NL} [pcm]	Δk_{eff} [pcm]
1	MOX-50%Th	−4167	−1112	+6649	+96	+1460
	SBU-50%Th	(−285%)	(−76%)	(+455%)	(+6%)	(+100%)
2	SBU-50%Th	−2754	+367	+655	+419	−1313
	WASB-50%Th	(−210%)	(+28%)	(+50%)	(+32%)	(−100%)
3	MOX-0%Th	−1753	+430	−1677	+9	−2992
	MOX-30%Th	(−59%)	(+14%)	(−56%)	(+1%)	(−100%)
4	MOX-30%Th	−7017	−1250	+6236	+171	−1857
	MOX-80%Th	(−378%)	(−67%)	(+336%)	(+9%)	(−100%)

Table 5. Contribution of each factor to variation of initial k_{eff}.

Th-232, which leads to the decrease of the reproduction factor by 1753 pcm. Moreover, because of harder neutron spectrum, the resonance escape probability decreases by 1677 pcm. As a result, the initial k_{eff} decreases by 2992 pcm and the required initial inventory of U-235 increases by 77.8 kg. If the thorium content further increases from 30 to 80%, as shown in the group 4, a large amount of U-238 is replaced by Th-232 in the reactor core. Because of a smaller resonance integral of Th-232 (85) compared to U-238's (275), the resonance escape probability (p) increases by 6236 pcm and contributes +336% but the reproduction factor decreases by 7017 pcm and contributes −378%. As a result, the initial k_{eff} decreases by 1857 pcm. Moreover, the required initial inventory of U-235 decreases by 206 kg because of a larger ACR.

3.4. Fuel cycle cost analysis

Using the levelized lifetime cost methodology described in Section 2.2.2, the fuel cycle cost of the four types of reactor cores (Th/U MOX, SBT, SBU, and WASB) analyzed in Sections 3.2 and 3.3 as a function of thorium content is shown in **Figure 16**. Compared with **Figures 11** or **13**, the fuel cycle cost changes with the same trend as the effective enrichment of U-235 or the initial inventory of U-235 in the reactor cores. When the thorium content is constant, the fuel cycle cost decreases with the increase of the spatial separation level. However, the difference of the SBU level and WASB level is small. The fuel cycle cost decreases with the increase of the

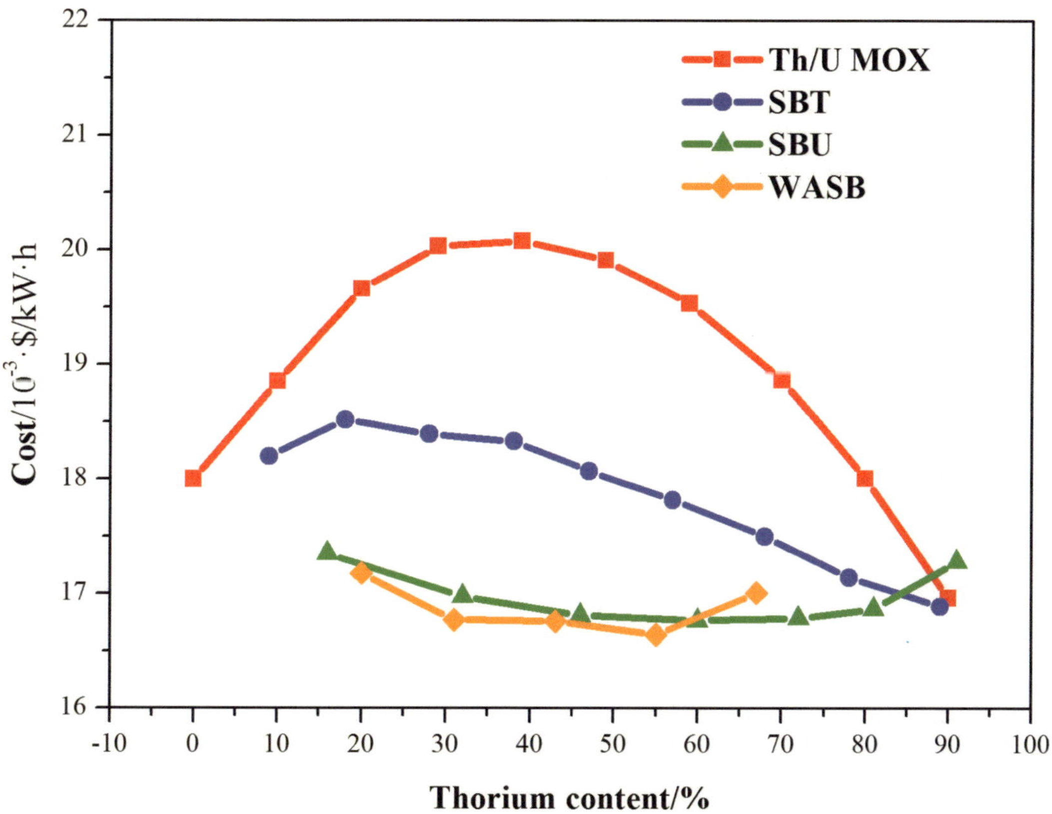

Figure 16. Fuel cycle cost of different spatial separation levels.

thorium content for the SBT, SBU, and WASB levels. However, it increases in the range of thorium content from 0 to 40% and decreases when the thorium content is larger than 40%.

The tight relationship between fuel cycle cost and initial inventory of U-235 mainly results from the composition of the cost. **Figure 17** presents the composition of fuel cycle cost for three different reactor cores, including natural uranium purchase, thorium purchase, uranium conversion, uranium enrichment, the fabrication of fuel blocks, and storage and the deposal of spent fuels. Because the required amount of natural uranium is 13 times of the inventory of heavy metal in the reactor core due to 0.7% U-235 in natural uranium, both the natural uranium purchase and uranium enrichment are 70% of total fuel cycle cost. The thorium purchase is only 2.5% because the thorium need not be enriched, and the required amount of thorium is by far less than the amount of the natural uranium. The fabrication cost is the highest unit price (777 $/kg HM) and the fabrication involves all heavy metals, so it is about 20% of the total cost. Although the unit price of the disposal of spent fuel is also high (610 $/kg HM), the disposal cost is only about 0.5% because of so-called time value.

The fuel cycle cost is mainly determined by natural uranium purchase (35–40%), uranium enrichment (32–35%), and the fabrication of fuel blocks (18–21%). The total of three items is 85–96% fuel cost. The fabrication cost is the same for all reactor cores because the inventory of heavy metal in the reactor core is the same. The natural uranium purchase and uranium enrichment are directly related to the initial inventory of U-235. The more is the inventory of U-235, the more are the required natural uranium and the uranium enrichment. As a result, the fuel cycle cost is the same trend as the initial inventory of U-235 or the initial effective enrichment.

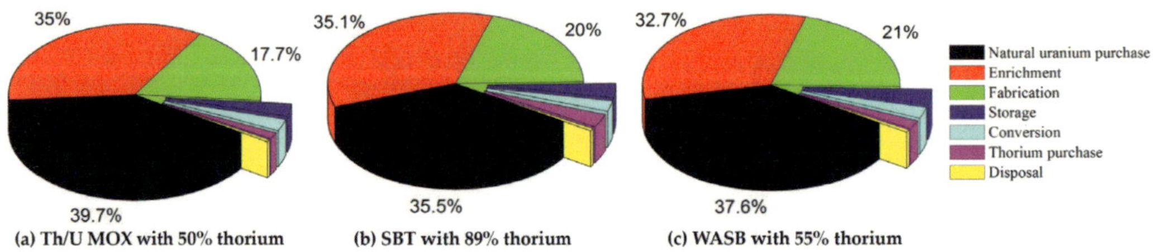

Figure 17. Composition of fuel cycle cost for some typical reactor cores.

4. Conclusions

In order to utilize thorium fuel in block-type HTRs with the features of inherent passive safety, high burnup, and hard neutron spectrum, two key factors of thorium content and spatial separation levels are chosen to be investigated. The thorium content represents the thorium/uranium fuel composition and the spatial separation level represents the spatial distribution of thorium/uranium fuels. For every thorium content,the spatial distribution of thorium and uranium fuels are concluded into four spatial separation levels, that is, no separation level (Th/U MOX), TRISO level (SBT), channel level (SBU), and block level (WASB) for the thorium-fueled block-type HTRs.

The nuclear performance of the reactor core, that is, initial effective multiplication factor, average conversion ratio, and initial inventory of U-235, under four spatial separation levels are obtained in the one-batch fixed-pattern refueling mode by the two-step calculation scheme developed based on the DRAGON. For every thorium content, the initial inventory of U-235 decreases with the increase of the spatial separation level from Th/U MOX to WASB, because spatial self-shielding effect is strengthened by the lumped thorium and uranium. However, the SBU level is nearly the same as the WASB level. If the multiple fuel batches and realistic refueling patterns are considered, the difference of the SBU from the WASB could be large. On the other hand, the initial inventory of U-235 decreases with the increase of the thorium content for the SBT, SBU, and WASB levels. However, for Th/U MOX level, it increases in the range of thorium content from 0 to 30% and decreases when it is larger than 30% because of the better performance of U-233 than Pu-239 in thermal reactors.

The performance difference of the four spatial separation levels is synthetically evaluated bythe levelized lifetime cost method.The fuel cycle cost of the Th/U MOX, SBT, SBU, and WASB changes with the same trend as the effective enrichment of U-235 or the initial inventory of U-235 in the reactor cores because the latter determines 70% of the total cost.

Acknowledgements

This work is supported by the National Natural Science Foundation of China (11405036).

Author details

Ming Ding[1]* and Jie Huang[1,2]

*Address all correspondence to: dingming@hrbeu.edu.cn

1 Harbin Engineering University, Harbin, China

2 China Nuclear Power Technology Research Institute, Shenzhen, China

References

[1] Baumer R, Barnert H, Baust E, Bergerfurth A, et al. AVR: Experimental High Temperature Reactor, 21 Years of Successful Operation for a Future Energy Technology. Dusseldorf: VDI-Verlag GmbH; 1990. p. 379

[2] Habush AL, Harris AM. 330-MW(e) fort St. Vrain high-temperature gascooled reactor. Nuclear Engineering and Design. 1968;7(4):312–321. DOI: 10.1016/0029-5493(68)90064-2

[3] Lung M, Gremm O. Perspectives of the thorium fuel cycle. Nuclear Engineering and Design. 1998;**180**(2):133–146. DOI: 10.1016/S0029-5493(97)00296-3

[4] International Atomic Energy Agency. Thorium Fuel Cycle-Potential Benefits and Challenges. Vienna: International Atomic Energy Agency; 2005. p 105

[5] Wang D. Optimization of a seed and blanket thorium-uranium fuel cycle for pressurized water reactors [thesis]. Cambridge: Massachusetts Institute of Technology; 2003.

[6] Shwageraus E, Zhao XF, Diriscoll MJ, Hejzlar P, Kazimi MS. Microhetero-geneous thoria-urania fuels for pressurized water reactors. Nuclear Technology. 2004;**147**,20–36

[7] Olson G, McCardell R, Illum D. Fuel summary report: shippingport lightwater breeder reactor (Technical report). INEEL/EXT-98-00799. Idaho Fall: Idaho National Engineering and Environmental Laboratory; 1999

[8] Radkowsky A, Galperin A. The nonproliferative light water thorium reactora new approach to light water reactor core technology. Nuclear Technology. 1998;**124**,215–222

[9] Ding M, Kloosterman JL. Thorium utilization in a small long-life HTR. Part II: Seed-and-blanket fuel blocks. Nuclear Engineering and Design. 2014;**267**,245–252

[10] Jie Huang, Ming Ding. Analysis of multi-scale spatial separation in a block-type thorium-loaded helium-cooled high-temperature reactor. Annals of Nuclear Energy. 2017;**101**:89–98

[11] Jie Huang, Ming Ding. Analysis of thorium content and spatial separation influence for seed and blanket fuel blocks in the AHTR. Progress in Nuclear Energy. 2016;**90**:182–189

[12] GA. Gas turbine-modular helium reactor (GT-MHR) conceptual design description report (Technical report). 910720. California: General Atomics; 1996

[13] Hébert A.. Toward DRAGON Version 4. In: Topical Meeting on Reactor Physics: Advances in Nuclear Analysis and Simulation (PHYSOR-2006). 10–14 September; Vancouver, Canada

[14] Hébert A. Collision Probability Analysis of the Double-Heterogeneity Problem. Nuclear Science and Engineering. 1993;**115**:177

[15] Sanchez R. Renormalized treatment of the double heterogeneity with the method of characteristics. In: Int. Mtg. on the Physics of Fuel Cycles and Advanced Nuclear Systems: Global Developments (PHYSOR-2004); 25–29 April 2004; Chicago, Illinois

[16] OECD. The Economics of the Nuclear Fuel Cycle (Technical report). Paris: Organization for Economic Co-operation and Development; 1994

[17] Holcomb DE, Peretz FJ, Qualls AL. Advanced high temperature reactor systems and economic analysis (Technical report). ORNL/TM-2011/364. Oak Ridge: Oak Ridge National Laboratory; 2011

[18] Jie H, Ming D, Yongyong Y. Fuel cycle cost analysis of Th/U MOX fuel in a block-type HTR. In: 8th International Topical Meeting on High Temperature Reactor Technology (HTR-2016); 6–10 November 2016; Las Vegas, USA

[19] Verrue J, Ding M, Kloosterman JL. Thorium utilisation in a small long-life HTR. Part III: Composite-compact fuel blocks. Nuclear Engineering and Design. 2014;**267**,253–262

[20] Duderstadt JJ, Hamilton LJ. Nuclear Reactor Analysis. New York: John Wiley & Sons, Inc.; 1976. pp. 295–299

[21] ORNL. Status of the physics and safety analyses for the liquid-salt-cooled very high-temperature (Technical report). ORNL/TM-2005/218. Oak Ridge: Oak Ridge National Laboratory; 2005

Section 4

Environmental Aspects of Power Plants

Chapter 7

Spatial Aspects of Environmental Impact of Power Plants

Boško Josimović and Saša Milijić

Additional information is available at the end of the chapter

http://dx.doi.org/10.5772/intechopen.68283

Abstract

Strategic Environmental Assessment (SEA) is one of the key instruments for implementing sustainable development strategies in planning in general, namely for analysing and assessing the spatial development concepts, in this case in the field of energy and planning of power plants. The SEA in energy sector planning has become a tool for considering the benefits and consequences of the proposed changes in space, also taking into account the capacity of space to sustain the implementation of the planned activities. This chapter examines the multi-criteria evaluation (MCE) method for carrying out an SEA for the power plants in Energy Sector Development Strategy of the Republic of Serbia (case study). The MCE method has found its use in the analysis and assessment of the energy sector spatial impacts on the environment and elements of sustainable development and, in this context, also considering the importance of impacts, spatial dispersion of impacts, their probability and frequency of occurrence, along with the elaboration of the obtained results in a specific, simple and unambiguous way. The chapter focuses on the consideration of aspects of environmental impact of all kinds of power plants, without taking into account the details regarding other aspects of energy sector development that are dealt with in the case study.

Keywords: strategic environmental assessment, power plants, multi-criteria evaluation

1. Introduction

The strategic environmental assessment (SEA) can be considered as the most important, the most general and the most comprehensive instrument for directing the strategic planning process towards the principles and objectives of environmental protection, as well as for making optimum decisions on future sustainable spatial development [1] can be considered one.

Since the very beginning of the last decade of the twentieth century until now, many authors [2–7] have written about the importance of the implementation of SEA in the process of making optimum decisions on sustainable spatial development and on the concept of development policy in different fields of social activities (energy, water resource management, infrastructure, tourism, etc.). The issue is therefore quite interesting, from both scientific and professional aspects, and is of great importance in creating any environmental policy.

The European Strategic Environmental Assessment Directive 2001/42/EC [8] prescribes the obligation to undertake SEA for plans, programmes and framework documents in different fields, thus also in the field of energy sector. By using the SEA, it is possible to consider the positive and negative implications of the proposed changes in the earliest stages of creating a development policy and accordingly direct the planning process in a way to increase positive impacts and minimize negative implications. The SEA process unavoidably implies the participation of the public in all stages of the planning process, thus emphasizing the contribution in decision-making, in this case in the field of the energy sector development [9–13]. Compared to other methods that contribute to decision-making, such as the traditional 'life cycle assessment' (LCA) [14–17] which is mainly used for the analysis of impacts in smaller territorial units, the SEA contributes to integrating the impacts at the strategic level of power plants planning (national, regional and, if necessary, international level). Given that the strategic level of planning requires a multi-dimensional consideration of phenomena and processes and that making appropriate decisions on spatial impacts and spatial development is a complex process, it is necessary to have an appropriate problem and methodological approach to making appropriate decisions on sustainability of offered concepts of development. This is particularly necessary in conceiving an energy sector policy, where the development tendencies can have significant implications for space and the environment. Multi-criteria analysis has been strongly advised by various authors with expertise in the energy sector [18, 19].

The chapter will present the possibility and importance of using the SEA for analysing spatial impact of power plants, regardless of whether it comes to fossil fuel power plants or power plants using renewable energy sources (RESs) because both can have strategically significant impacts (positive and negative) on the space where they are built. The chapter will examine the multi-criteria evaluation (MCE) method for carrying out an SEA for the case study–Energy Sector Development Strategy of the Republic of Serbia by 2025 with projections until 2030 [20] (herein after referred to as the 'Strategy').

2. Methodological framework

The concept of the SEA methodologies, unlike the diverse, precise and highly operable tools used in environmental engineering or other science-based areas, is rather fuzzy [21]. Some authors [22–24] believe that there is no uniform methodological approach to the SEA process because its use in very thematically different planning processes is not appropriate, but it is the most appropriate to use different methodological approaches or their combination that would be aligned with specific conditions in which the planning process is carried out [22, 25]. In accordance with the above-mentioned, it can be concluded that the SEA process should be based on an interdisciplinary approach and open to the use of different methodological approaches. Such flexibility in the approach to the SEA process leaves the possibility of finding the best solutions in accordance with specific approaches to the planning process. Generally speaking, the SEA techniques and methodologies derive from the traditional Environmental Impact Assessment (EIA) and policy appraisal/plan evaluation studies [25, 26], ensuring that methodologies would not become a barrier for institutional promotion of the SEA [27]. A variety of possible techniques for conducting the different steps of SEA have been further analysed and discussed by others [1, 25, 27–29]. In addition, Marsden [30] pointed out that, in terms of methodologies, the SEA process is dominantly based on the qualitative or semi-quantitative approach which to a great extent depends on the skills and knowledge of experts involved in the evaluation process. Such an approach also implies a certain degree of subjectivity that should be minimized by using different software packages (like geographic information system [GIS]) in combination with their adequate use experiences accumulated through comparative studies of past schemes and applications [21].

The procedural and methodological framework for SEA is shown in **Figure 1**. The SEA process begins with decision-making on undertaking the SEA, as well as on its scope and contents. This stage involves the defining of framework for investigations to be carried out in the SEA process along with the unavoidable participation of the public, relevant institutions and non-governmental organizations (NGOs). The next stage includes the analytical part of the SEA process implying the analysis of the state of the environment in the area under investigation using the GIS technologies [31–33]; analysis of the strategic concept of development (in this case in the energy sector); comparative analysis of the planning and strategic documents relating to the space that is subject of investigation and to the specific field of investigation; other investigations and analyses of importance for the specific field of investigation. The next stage included the setting of the SEA objectives, relevant indicators (**Table 1**) and evaluation criteria (**Tables 2** and **3**), followed by an impact assessment procedure in which the first stage included the evaluation of alternative scenarios and the selection of the most suitable alternative. Then, the process of multi-criteria evaluation (a semi-quantitative method) followed, representing the focal point of this chapter (presented in point 3 herein). The role of multi-criteria evaluation is to identify the influence of the activities planned in the space in which they are being undertaken (the prediction of spatial influences) according to the SEA objectives. When the impacts of the Strategy are identified in such a way, then it is possible to elaborate and present them in a simple and unambiguous way and in a way that is clear to all actors in the SEA process, namely the actors included in the decision-making process.

The specific SEA objectives (**Table 1**) were set in certain fields of environmental protection. The specific SEA objectives are the concrete, partially qualified statements in a form of guidelines and actions (measures, works and activities) for the implementation of these changes. The

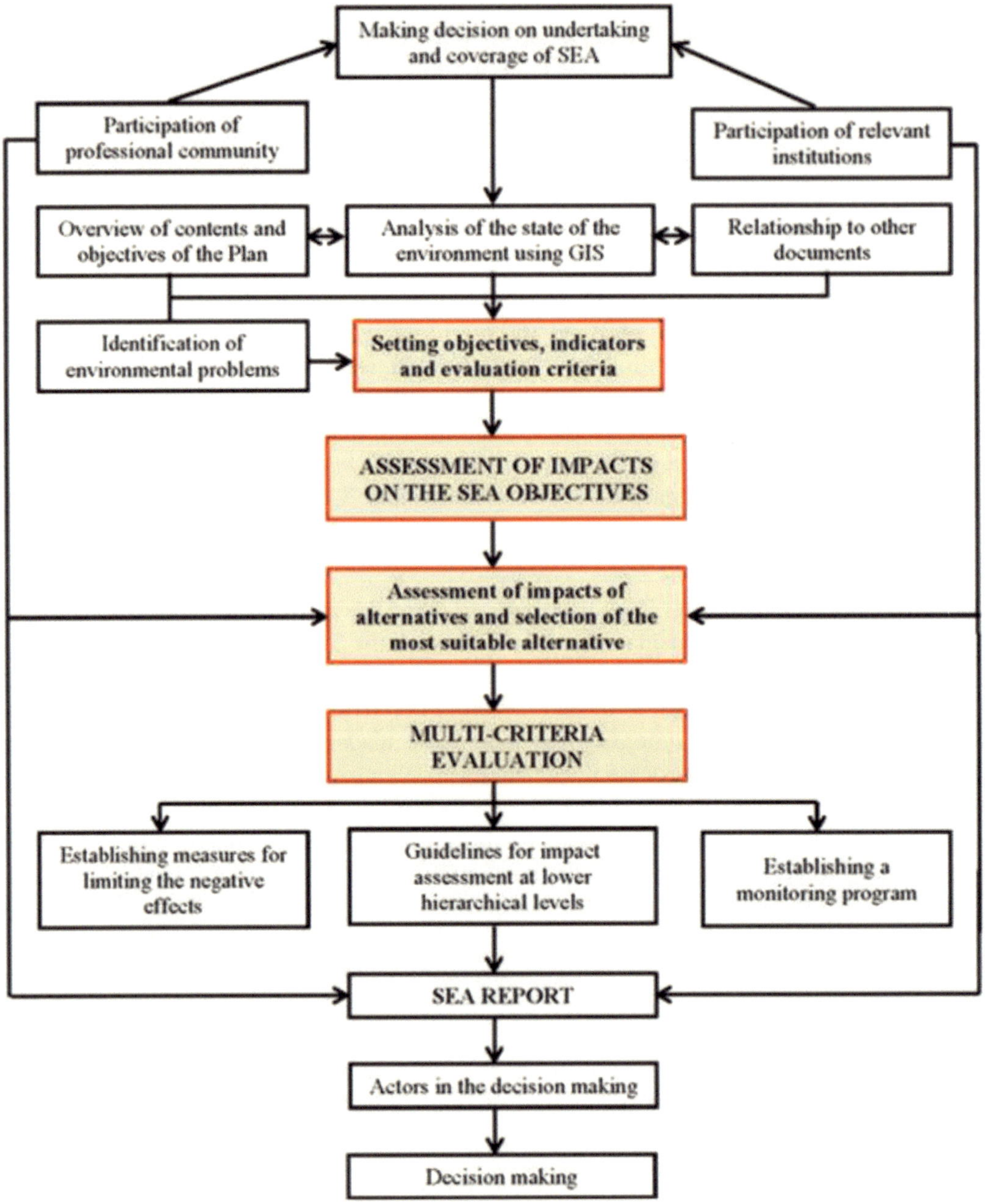

Figure 1. Procedure and methodological framework for the SEA [11].

specific SEA objectives are primarily a methodological measure through which the effects of power plants (and energy sector in general) on the environment are identified and checked. The multi-criteria evaluation of the planning solutions is carried out in relation to these objectives to obtain a clear idea about possible effects of the planning process and make optimum decisions on future sustainable spatial development accordingly.

http://dx.doi.org/10.5772/intechopen.68283

Specific SEA objectives	Indicators
AIR	
- Reducing air pollutant emissions to prescribed emission levels	- Particle emissions, and SO_2 and NOx emissions - Frequency of exceeding the daily limit values for soot, SO_2 and NO_2 - Changes in greenhouse gas emissions, primarily in CO_2 (%) - Increase of RES share in energy balance (%)
WATER	
- Reducing surface and groundwater pollution to the level that will not affect their quality - Mitigating negative effects of energy facilities on hydrological regimes	- BOC and COC in watercourses affected by energy facilities and activities - Temperature changes in watercourses - Changes in water regimes - Changes in water quality class (%) - Reused and recycled water as a result of energy sector activities (m^3)
SOIL	
- Protection of agricultural and forestry land - Reducing soil degradation and erosion	- Changes in forest cover (%) - Changes in agricultural land area (%) - Share of degraded areas as a result of activities in the field of energy (%) - Land area threatened by soil erosion processes (ha)
NATURAL VALUES	
Specific SEA objectives	**Indicators**
- Landscape protection - Protection of natural values and areas - Biodiversity conservation – avoid irreversible losses	- Share of re-cultivated areas in the overall area of degraded regions (%) - Number of energy facilities that may cause landscape changes - The area of all protected natural areas that may be affected - Number of endangered flora and fauna that may be affected
CULTURAL AND HISTORIC HERITAGE	
- Cultural properties protection	- Number and importance of protected immoveable cultural properties that may be affected
WASTE	
- Improving waste utilization, treatment and disposal	- Total annual amount of waste generated in energy sector (t) - % of total amount of waste subject to re-use, recycling and treatment
POPULATION HEALTH	
- Reducing negative effects of energy industry on human health	- Frequency of respiratory diseases (%) close to energy facilities - Frequency of diseases which can be associated with energy activities - Number of people affected by noise generated by energy facilities
SOCIAL DEVELOPMENT	
- Better quality of life - Preservation of population density in rural areas	- Improvement of energy efficiency in residential buildings (%) - Number of households displaced as a result of energy activities
INSTITUTIONAL DEVELOPMENT	
- Improve environmental protection and management and control services	- Number of measuring points in monitoring system
ECONOMIC DEVELOPMENT	
- Encouraging economic development - Promote local employment - Reducing dependency on sources of imported energy - Reducing transboundary negative environmental impacts	- % of employed in the energy sector with salary above the average salary - Reduction in number of unemployed as a result of growth in the energy sector employment (%) - Number of environmental protection programs for energy sector development - Number of IPPC energy facilities with transboundary impacts
NATURAL RESOURCES	
- Rational use of non RES and increasing the use of RES - Improving energy efficiency - Introducing cleaner technologies	- Final energy consumption per capita - Share of renewable energy sources in total energy use - Improvements in energy efficiency (% of reduction in energy use)

Table 1. SEA objectives and indicators for the Strategy [34].

The sustainable development indicators (**Table 1**) are needed to identify trends of moving towards or away from sustainability, as well as to set goals for improving general well-

Impact magnitude	Designation	Description
Critical	-3	Significant environmental overload
Greater	-2	Environmental disturbance of great extent
Smaller	-1	Environmental disturbance of smaller extent
No impact	0	No direct and/or unclear environmental impact
Positive	+1	Smaller positive environmental changes
Favourable	+2	Favourable environmental changes
Very favourable	+3	Changes that significantly improve the quality of life
Impact significance	**Designation**	**Description**
International	I	Possible transboundary impact
National	N	Possible impact at the national level
Regional	R	Possible impact at the regional level
Local	L	Possible impact of local character
Probability	**Designation**	**Description**
100%	S	Impact will definitely occur
More than 50%	L	Likely impact
Less than 50%	P	Possible impact

Table 2. SEA evaluation criteria for the Strategy [34].

being. In 2008, the Republic of Serbia adopted the National Energy Sector Development Strategy [35] which contains principles and priorities in sustainable development, as well as 76 indicators for tracking the progress of Serbia towards sustainable development. These indicators have been selected from the set of UN indicators, but not all of indicators are used in Serbia. The indicators are specified in the Law on Spatial Plan of the Republic of Serbia [36]. The Regulation on the National List of Environmental Indicators [37] prescribes the list of environmental indicators, which have been used herein. The SEA indicators have been selected in accordance with the above-mentioned SEA objectives. This set of indicators is based on the 'cause-effect-response' concept. The indicators of cause denote human activities, processes and relationships affecting the environment, the indicators of effect denote the state of the environment, while the indicators of response define strategic options and other responses aimed at changing 'consequences' for the environment.

Based on an analysis of the possibility of primarily considering the spatial aspect, as well as the problematic aspect of potential impacts, three sets of criteria with a total of 14 individual criteria were defined. The criteria used in the MCE of the planning solutions were related to the magnitude (intensity) of the impact, the spatial dimension of the impact and the impact probability (**Table 2**).

The importance of identified impacts for achieving the SEA objectives is evaluated. The impacts of importance for the subject Strategy are those which have stronger or greater effects on the international (transboundary), national or regional level, according to the criteria shown in **Table 3**.

The methodological framework for the SEA presented in this chapter is centred on a plan-based approach to and the use of MCE method for the planned activities and strategic determinants in

Level	Impact magnitude		Designation of significant impacts
International level: I	Strong positive impact	+3	I+3
	Greater positive impact	+2	I+2
	Strong negative impact	- 3	I-3
	Greater negative impact	- 2	I-2
National level: N	Strong positive impact	+3	N+3
	Greater positive impact	+2	N+2
	Strong negative impact	- 3	N-3
	Greater negative impact	- 2	N-2
Regional level: R	Strong positive impact	+3	R+3
	Greater positive impact	+2	R+2
	Strong negative impact	- 3	R-3
	Greater negative impact	- 2	R-2

Table 3. SEA criteria for evaluating strategically significant impacts of the Strategy [34].

relation to the capacity of space as a basis for the valorization of space earmarked for sustainable development [38].

The development of the MCE method is linked to the early 1970s of the twentieth century when many authors [39–41] began to develop such an approach. When first developed, the MCE was characterized by the methodological principle of multi-criteria decision-making (MCDM) with little or no participatory mechanisms included [42, 43]. The initial idea was to elicit clear preferences from a decision-maker and then solve a well-structured problem by means of mathematical algorithms (e.g. to design an engine by taking into account its power, weight and efficiency). Over time, this ambitious idea has been directed towards a more rational approach [44] and to a constructive or a creative approach [45], which has led to an approach based on the development of the multi-criteria decision aid (MCDA) that is characterized by placing in focus the process of decision-making on specific development processes, namely the raising of the level and quality of this process along with permanently including the public into all stages of creating the future development [46–48]. This approach has resulted in the emergence of the participatory multi-criteria evaluation (PMCE) [46, 49] and social multi-criteria evaluation (SMCE) [50]. Nowadays, the MCE method is often recommended as a convenient support in the decision-making process because of its capacity to point out in many ways multiple alternatives of development on the basis of assessing criteria related to the environment and socio-economic aspects of sustainable development [51–54].

The MCE method presented and elaborated in this chapter was originally defined in a scientific research project entitled 'Method for Strategic Environmental Assessment in Planning' (2005–2007). The method was later developed and upgraded through several domestic and international scientific projects, in which the Institute of Architecture and Urban & Spatial Planning of Serbia (IAUS) also participated. The results of the research and method development have found their applicability in the drawing up of a number of strategic documents at national and regional levels, in the fields of energy, water resource management, waste management, tourism and so on. An example of the application of the MCE method is shown in the text that follows.

3. Case study: MCE method in the SEA for the power plants in the Strategy

The results presented in this part of the chapter have been taken from the SEA for the Strategy [34] (case study), conducted by the authors of this chapter. The mentioned methodological approach was used in drawing up the mentioned document, while only the results that are in the context of the theme of this research and that were partially slightly modified to adapt them to the format of this chapter are elaborated herein.

3.1. Subject of the Strategy and SEA

The approach of integrated and continuous planning was used in creating the Strategy and carrying out the SEA with a focus on finding the measures for sustainability through integrating the realistic goals and the potentials in the field of energy, on the one hand, and the goals of and needs for environmental protection, quality of life of population and socio-economic development, on the other hand. The Strategy will be a framework for the development of the energy system of the Republic of Serbia with all possible (positive and negative) implications for the environmental quality. The purpose of carrying out the SEA for the Strategy is to consider possible negative effects on/trends of environmental quality, as well as to develop the guidelines for their reduction, namely bringing them within acceptable limits without creating the conflicts in space and taking into account the environmental capacity in the considered area.

The SEA is not focused only on the analysis of strategic commitments that can imply negative impacts and trends but also on the strategic commitments that can contribute to environmental protection and quality of life of population. In this context, the potential environmental impacts of planned activities were analysed in the SEA and were evaluated relative to the objectives and indicators.

The chapter of the Strategy related to the priorities of the energy sector development mentions the power plants as one of the keys and essential potentials for development, but also as a possible cause of significant problems in the environment. For this reason, the chapter emphasizes that there is the need to introduce the latest technologies in energy generation, both in the thermal power plants and in the field of renewable energy sources (RES) for which there are significant potentials, but they have not been sufficiently used in the Republic of Serbia.

Although other important strategic priorities in the field of energy are also dealt with in the Strategy, due to the fact that power plants have significant environmental impacts on the space in which they are built, the SEA gives special attention to this aspect, and exactly this aspect is elaborated further herein.

3.2. Evaluation of power plant impacts using the MCE method

The SEA for the Strategy singles out the total of 29 planned activities/strategic priorities that was included in the multi-criteria evaluation process. Nine out of 29 strategic priorities relate to the projects of power plants (**Table 4**). However, save for the locations defined by the document entitled 'Cadaster for Small Hydro Power Plants in Serbia' [55], the Strategy does not determine micro-locations for the majority of power plants (this particularly refers to the power plants using RES), so the evaluation could be based on predictions and on keeping pace with trends in space on the basis of micro-location determination and knowledge of general regularities and potential environmental impacts that certain types of power plants imply.

Table 4 shows the strategic priorities in the Strategy, which directly relate to the projects of power plants and which were also included in the multi-criteria evaluation presented further herein. In addition to these priorities, the Strategy also formulates priorities that directly relate to power plants because they are actually in their function. This includes, for example, the extension of coal open-pit-mining areas and opening of new coal open-pit mines which supply thermal power plants with coal. These strategic priorities were also elaborated in the SEA for the Strategy, but are not presented herein.

Each individual strategic priority in the Strategy relating to the projects of power plants (**Table 4**) was included in the multi-criteria evaluation process by forming matrices (**Table 5**) in which the mentioned priorities relating to power plants 'intersect' with the SEA objectives and indicators (**Table 1**). They were evaluated according to the adopted groups of criteria (**Table 2**).

The matrices were formed for the first two groups of criteria (impact magnitude and impact significance) because it is possible to identify the impacts of strategic importance already based on these two groups of criteria, namely the impacts where evaluation is within the values adopted and presented in **Table 3**.

The evaluation of strategic priorities was semi-quantitative, qualitative and aligned with the level of detail/generality that characterizes the strategic documents such as the Strategy. It was based on the planning approach through which the spatial (territorial) impacts were considered, as well as on possible conflicts in space that can occur in the interaction between the existing and the planned activities in a specific space without going into the technical and technological aspects of potential impacts that are not possible to be identified at this level of planning.

Each individual strategic priority was then also presented in the form of graph (**Figure 2**) and in a way comprehensible to the general public and decision-makers, this being in accordance with the Arhus Convention principles.

The evaluation of strategic priorities in the Strategy that relate to power plants, and which were elaborated in the SEA, is presented further herein.

A table was formed based on the adopted criteria for determining the strategically significant impacts (**Table 3**) and results of the multi-criteria evaluation presented in the form of matrices

Sector of the Strategy	Priority activities
Electric power system	Reconstruction and withdrawal of thermal power plants pursuant to Large Combustion Plants Directive
	Coal-fired power plants of capacity of 700MW by 2025 (350MW by 2020)
	"Bistrica" reversible HPP and/or "Djerdap 3" reversible HPP
	Construction of combined heat and PPs of capacity of approx. 450 MW by 2020
Renewable energy sources	Development of wind farms
	Development of hydroelectric power plants and small hydro power plants
	Development of biomass power plants
	Development of solar power plants
	Development of geothermal power plants

Table 4. Strategic priorities in the strategy included in the SEA relating to the projects of power plants [34].

(**Table 5**). The table identifies the strategic priorities that imply strategically significant impacts (positive and negative) on the environmental quality and elements of sustainable development, namely on the SEA goals, and determine the rank of impacts identified in such a way (**Table 6**).

ASSESSMENT OF IMPACT MAGNITUDE																					
Strategic priorities	SEA objectives																				
	1	2	3	4	5	6	7	8	9	10	11	12	13	14	15	16	17	18	19	20	21
Reconstruction and withdrawal of thermal power pursuant to the Large Combustion Plants Directive	+3	+3	0	+2	0	0	0	+1	0	+1	+1	+1	+1	0	+1	0	0	+1	-1	+1	+3
Coal-fired thermal of capacity of 700MW by 2025	0	0	0	-1	0	-2	0	-1	0	+1	0	+1	-1	0	+2	+2	+2	0	-1	0	+2
"Bistrica" reversible HPP and/or "Djerdap 3" reversible HPP	0	0	-3	-2	-1	-2	-1	-2	0	0	0	+1	-1	0	+2	+2	+3	-2	+3	0	+3
Development of combined heat and power plants using gas engines of capacity of approx. 450 MW by 2020	+1	0	0	0	0	-1	0	0	0	0	0	+1	0	0	+2	+1	0	0	+2	0	+1
Development of wind farms	+3	0	-2	-1	0	-1	-1	-2	0	+2	+1	+1	0	0	+1	+1	+1	-1	+3	0	+2
Development of HPPs and SHPPs	+3	0	-2	-1	0	-1	-1	-2	0	+2	+1	+1	0	0	+1	+1	+1	-1	+3	0	+2
Development of biomass power plants	-1	0	0	0	0	0	0	0	0	+1	+1	+1	0	0	+1	+1	+1	0	+3	0	+1
Development of solar power plants	+3	0	0	-1	-1	-1	0	0	0	0	+1	+1	0	0	+1	+1	+1	0	+3	0	+3
ASSESSMENT OF IMPACT SIGNIFICANCE																					
Strategic priorities	SEA objectives																				
Reconstruction and withdrawal of thermal power pursuant to the Large Combustion Plants Directive	R	R		L				L		L	R	L	L		L			I	L	R	R
Coal-fired thermal power plants of capacity of 700MW by 2025				L		L		L		L		R	L		N	L	N		N		R
"Bistrica" reversible hydropower plant and/or "Djerdap 3" reversible HPP			I	I	I	I	I	N				I	L		N	L	N	I	N		N
Development of combined heat and power plant using gas engines of capacity of approx. 450 MW by 2020	L					L						L			L	L			L		L
Development of wind farms	N		I	L		L	L	I		L	L	N			N	N	N	I	N		N
Development of HPPs and SHPPs	N		I	L		L	L	I		L	L	N			N	N	N	I	N		N
Development of biomass power plants	L									L	L	L			L	L	L		L		L
Development of solar power plants	L			L	L	L					L	L			L	L	L		L		L

[*]Criteria according to **Table 2**.

Table 5. Matrices of spatial impacts of power plants on the environment and sustainable development [34].

http://dx.doi.org/10.5772/intechopen.68283

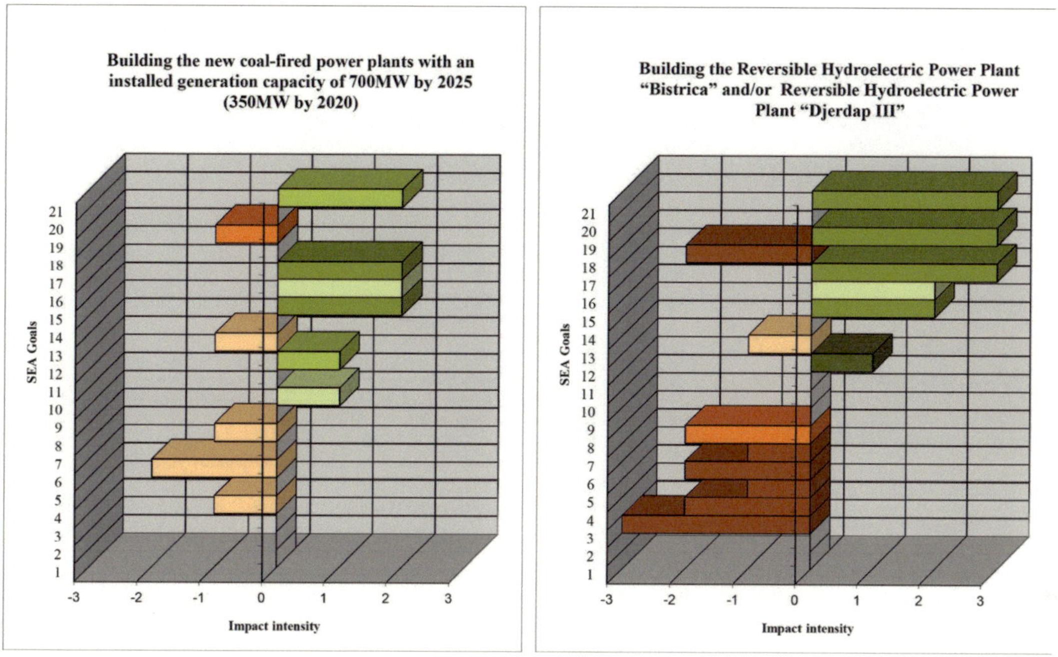

Figure 2. Examples of the presentation of obtained results in the form of graph [34].

3.3. Summary of significant strategic impacts

On the basis of the evaluation of impact significance shown in **Table 6**, it can be concluded that the Strategy produces significant number of strategically important, both positive and negative, environmental impacts. This is not at all surprising considering that the main idea of the Strategy was to create a concept of development of an energy sector which would rely in all of its segments on the principles of the prevention and active protection of the environment, its main factors, thus also on the improvement of human health by primarily permanently reducing the exposure to air pollution resulting from the operation of thermal power plants operating with inadequate technologies.

The identified negative impacts of the Strategy are the result of development in the energy sector which, in the Republic of Serbia like in other countries, dominantly relies on the existing energy potentials. The most important negative impacts are the result of thermal power plant operations, namely the result of the surface coal exploitation necessary for their operation. In this context, the impacts of the main environmental factors (air, water, soil and biodiversity), as well as impacts on landscape, stand out by their importance. The impacts on social factors as the result of extension of open-pit mines, due to which parts or even entire settlements are moved to completely new location, are of special importance.

Most of the mentioned impacts are not significantly spatially dispersed, but, nevertheless, they were assessed as being of strategic importance due to the assessed impact intensity, as shown in **Table 6**.

Strategic priorities	Identification and evaluation of significant impacts		Reasons explanation
	SEA objective	Rank	
ENERGY SYSTEM			
Reconstruction of thermal power plants pursuant to the Large Combustion Plants Directive	1	R+3 / S	It is expected that people's quality of life will be significantly improved at the regional level by reducing emissions of ambient air pollutants and reducing negative impacts on water quality. The introduction of cutting edge technologies in thermal power plants will contribute to this.
	2	R+3 / L	
	21	R+3 / S	
Development of new coal-fired thermal power plants of capacity of 700MW by 2025 (350MW by 2020)	15	N+3 / P	Significant positive effects are expected, which will spur economic growth and reduce dependence on energy imports by using new technologies in production processes.
	17	N+2 / S	
	21	R+2 / S	
Construction of the "Bistrica" reversible hydropower plant and/or "Djerdap 3" reversible hydropower plant	3	I-3 / S	There will definitely be negative transboundary impacts on hydrological regime of watercourses on which hydropower plant will be built, which may also cause the loss of forest and agricultural lands upstream of water intake (dam). Construction of the hydropower plant could have visual impact on landscape quality, but also impact on biodiversity. Positive impacts relate to spurring economic growth, reduction in dependence on energy imports, increase in the share of power produced from renewable energy, and introduction of clean technologies in electricity production.
	4	I-2 / L	
	6	I-2 / S	
	8	N-2 / P	
	15	N+2 / P	
	17	N+3 / S	
	18	I-2 / S	
	19	N+3 / S	
	21	N+3 / S	
RENEWABLE ENERGY SOURCES			
Increase in the share of power produced from renewable energy	1	N+3 / L	The expected positive effects will contribute to reducing harmful air pollution by increasing renewable energy use, i.e. by introducing clean technologies in electricity production processes. Certain projects can have negative effects on certain natural resources (effects of small hydropower plants on water resources) and biodiversity (effects of wind farms on ornithological fauna and chiropters).
	3	I-2 / L	
	8	I-2 / P	
	19	N+3 / S	
	21	N+2 / S	
Development of HPPs and SHPPs	1	R+3 / Lk	The expected positive effects will contribute to reducing harmful air pollution by increasing renewable energy use, i.e. by introducing clean technologies in electricity production processes. The negative impacts on the hydrological regime of watercourses on which they will be built are likely to happen. Hydroelectric power plants. These impacts will be particularly pronounced during the construction of HPPs with powerhouse at the toe of the dam and during the construction of SHPPs as a result of creating an artificial lake upstream from the dam, as well as in cases when several HPPs and SHPPs are planned on the same watercourse (cumulative and synergetic impacts). The positive impacts of national character that relate to the examples of clean technologies, renewable energy sources and reduced energy dependence by using significant hydro-energy potentials are certain to happen.
	3	I-2 / L	
	8	I-2 / P	
	19	N+3 / S	
	21	N+2 / S	
	3	R-2 / P	
	9	N-3 / P	
	17	N+2 / S	
	19	N-2 / P	
	21	N-2 / P	

Table 6. Identification and evaluation of strategically significant impacts with explanations [34].

Certain negative implications are also expected due to the construction of the 'Bistrica' HPP and/or construction of the 'Djerdap 3' reversible HPP, which would have negative effects on hydrological regime of watercourses on which their construction is planned, as well as on biodiversity and ichthyofauna, and will cause possible changes in the use of agricultural and forest lands.

By accepting the provisions of the Espoo Convention [56] and Kiev Protocol [57], the Republic of Serbia is obligated to notify the neighbouring countries on the projects that could have transboundary impacts. According to the Espoo Convention on Environmental Impact Assessment in a Transboundary Context (Espoo Convention), the transboundary impacts are defined as all impacts (positive and negative) of a certain activity or a project that affect the environment under the jurisdiction of another country. The implementation of the Espoo Convention provisions implies that the relevant institutions of the neighbouring countries should provide information on the planned activities and their environmental impacts so that the affected country could participate in making optimum decisions on planning and future realization of such activities.

In addition to the projects defined by the Strategy that could have transboundary impacts, such as the RHPP 'Bistrica' and/or RHPP 'Djerdap 3' projects, the transboundary environmental impacts can also occur as the result of the realization of the following projects:

- Wind farms—possible transboundary impacts on the flying fauna (ornithological fauna and chiropters) on the border between Serbia and Romania (actually, all wind farm projects planned to be built in the north-eastern part of the Republic of Serbia);
- Small hydropower plants on transboundary watercourses (actually, the watercourses flowing from the territory of the Republic of Serbia into the territory of the neighbouring countries)—possible adverse impacts on benthonic organisms and ichthyofauna on boundaries with Montenegro, B&H and Romania;
- Coal-fired power plant projects—possible impacts on the air quality and international rivers.

Other planed activities that can imply the transboundary impacts of strategic importance have not been identified.

In addition to the adverse impacts, by using the MCE method in carrying out the SEA for the Strategy, a whole series of strategically significant positive impacts were also identified:

- Environmental quality: positive impacts on the quality of main environmental factors (air, water and soil) and reduction of greenhouse gas (GHG) emissions by increasing the use of RES and the use of clean technologies in thermal power plants in accordance with Directive 2001/80/EC on the limitation of emissions of certain pollutants into the air from large combustion plants and with the Directive 2010/75/EU on industrial emissions for new projects; withdrawal of all thermal power plants with generating capacity less than 300 MW; the implementation of a whole set of energy-efficiency measures will contribute to more rational energy use, that is, reducing the production of the required amount of energy for the same amount of required energy; the improvement of legal regulations in the area of energy and their harmonization with the EU regulations; the development of institutions for the implementation of improved legal regulations related to the environmental pollution reduction and main environmental factors;
- Socio-economic development: Improvement of the energy sector as a driver of economic growth and development, the market-oriented formation of energy prices and prices for energy-generating products; the development of domestic industry and commercial

scientific research sector for the transfer of leading edge technologies in the field of energy; strict implementation of energy-efficiency measures in final energy consumption; labour market mobility; as well as overall energy sector development, will represent a long-term contribution to sustainable economic development and rational use of non-renewable energy sources, that is, increasing the share of energy from RES.

4. Discussion and conclusions

The Strategy represents a strategic framework for implementing the policies and measures in the field of energy, also including the aspect of using the energy potentials by planning and building new power plants or by modernizing the existing ones. The possible implications for the environment as a result of the realization of such projects, as well as the significant public participation (NGO, population, expert groups, relevant institutions, investors and public companies in the field of energy) in the decision-making processes, undoubtedly indicate the need to consider this aspect in creating the policies in the field of energy, either at the national, regional or at the local level. The reason for the interest of the public in power plants, and energy sector in general, lies in the nature of their operation and possible significant spatial/territorial impacts. In this context, besides different studies on environmental protection carried out at the level of specific investment projects for individual power plants (e.g. EIA or LCA), it is of particular importance to carry out an environmental impact assessment at the strategic level of planning which should direct a strategic-planning document to the goals of sustainability and prevent potential conflicts in space in the earliest stage of creating an energy policy. The SEA is exactly the instrument that meets these specific requirements. This has been much written about in scientific literature, which is briefly summarized in the 'Introduction' section of this chapter.

The Energy Strategy [20] is a specific document due to the fact that it establishes a completely new and altered concept of energy sector development relative to the existing one that has been, at least in the Republic of Serbia, assessed as unsustainable and ecologically unacceptable. In addition, the Strategy gives a special attention to power plants (the existing and the planned ones) that are particularly significant for analysis from the aspect of environmental protection. This conditioned the specificity of the subject SEA within which it was necessary to make a symbiosis of all existing phenomena and the processes in space and predictions about potential impacts of power plants on the environment. This served for defining the SEA objectives and corresponding indicators, but also the criteria, based on which the strategic priorities related to power plants were assessed. By using the multi-criteria evaluation and semi-quantitative methods according to three groups of criteria, as well as by presenting the results using the matrices and graphs, the results were shown in a clear and unambiguous way for each of the strategic priorities. Different aspects of impacts (significance, spatial extent and probability) were encompassed. The results of the used MCE method were taken as a basis for defining the appropriate measures/guidelines for the environmental protection and monitoring which should be used in implementing the Strategy, namely in determining the micro-locations for new power plants and in preparing the investment projects. These measures/guidelines defined the SEA by each type of power plant, particularly taking into account new power plant projects and instructions for their realization in a way that provides the preventive protection of space and the environment, thus enabling all actors in the process of creating and adopting the Strategy to make conclusions on positive and negative implications of all identified impacts in a simple way, while the SEA in this process also serves as a regulatory and control tool for all future investments in the energy sector.

Unlike some SEA that are oriented towards identifying 'for and against' alternatives, in the SEA for the Strategy it was relatively simple to propose to actors to make decisions on which solution is the most suitable for the implementation of the Strategy because positive changes brought by its implementation are many times greater relative to the possible negative implications. This was affected by the methodological approach that is conceptually conceivable; clear way of presenting the obtained results which enables great public participation in critical stages of the SEA (identification of possible positive and negative impacts of the planning concepts); the use of semi-quantitative methods in assessing the impacts and defining the appropriate guidelines for the environmental protection and monitoring.

The specificity of the presented approach is reflected in the identification of the objectives and indicators of the SEA that is based on the analysis of a complex symbiosis of environmental quality, strategic frameworks defined in different strategic and planning documents and the Strategy. The objectives and indicators obtained through this procedure are a good basis for assessing the complex implications of the planned activities for the space and possible mutual interactions between different sectorial determinations regarding the elements of sustainable development. A clear presentation of obtained results of multi-criteria evaluation in the form of matrices, and particularly in the form of graphs, is particularly important in the SEA stages involving public participation. However, the use of semi-quantitative method of expert decision-making brings a certain dose of subjectivity, which can be tentatively regarded as a lack of such an approach. On the other hand, the use of different mathematical methods like ARAS (additive ratio assessment) [58] or AHP (analytical hierarchy process) [59] at the level of strategic planning and management is not necessary, and often impossible because of a lack of appropriate inputs. In this case, for example, there was no determination of micro-locations for the planned power plants that will use renewable energy sources, so the prediction of impacts could be only made based on the knowledge on energy potentials, for example, the energy to be obtained from renewable sources at the micro-location level, the state of the environment on these micro-locations and potential capacities (installed power) in these potentially suitable micro-locations. Such obtained results of the assessment of spatial/territorial impacts of strategic priorities represent exactly a good basis for establishing adequate guidelines and for the use of the above-mentioned mathematical methods at a lower level of impact assessment, namely in carrying out the EIA, which is actually also an obligation in carrying out the SEA. In this context, this lack should be understood tentatively, but should in no way be ignored in the process of carrying out the SEA and in making appropriate decisions.

The guidelines for environmental protection were established based on the results of multi-criteria evaluation of strategic priorities that relate to the planning, building or modernization of power plants. The guidelines were given for the most significant facilities/projects envisaged in the Strategy, namely for the energy facilities/projects which can be significant polluters by nature of their operation.

A special contribution of the SEA process was laid in the identification of possible transboundary impacts and in the establishment of the transborder cooperation with the neighbouring countries. In this context, the cooperation has been established with the Republic of Croatia, Romania and Republic of Bulgaria, which have showed their interest including themselves into the SEA process and in officially submitting their views regarding the identified possible transboundary impacts. Their views were adequately taken into account in the preparation of the SEA Final Report.

Performing the SEA for the needs of the Strategy resulted in making appropriate decisions in the process of establishing sustainable energy sector development, while the implementation of documentation in practice should contribute to the applicability of results and to the possible monitoring of SEA contribution in the process of planning both power plants and all other plans in the energy sector.

Acknowledgements

This chapter has resulted from the research conducted within the scientific projects 'Spatial, ecological, energy, as well as social aspects of settlement development and climate change – their mutual impacts', TR 36035; and 'Sustainable spatial development of Danubian Serbia' TR36016; all financed by the Ministry of Education, Science and Technological Development of the Republic of Serbia during the period 2011–2017.

Author details

Boško Josimović* and Saša Milijić

*Address all correspondence to: bosko@iaus.ac.rs

Institute of Architecture and Urban & Spatial Planning of Serbia, Belgrade, Serbia

References

[1] Sadler B, Verheem R. Strategic environmental assessment: Key issues emerging from recent practice. Hague 7, Ministry of Housing, Spatial Planning and the Environment; 1996

[2] Maričić T, Josimović B. Overview of strategic environment assessment/sea/system in southeast Europe. Architecture and Urbanism. 2005;**16-17**:66-74

[3] Nilssona M, Björklundb A, Finnveden G, Johanssonc J. Testing a SEA methodology for the energy sector: A waste incineration tax proposal. Environmental Impact Assessment Review. 2005;**25**:1-32. DOI: 10.1016/j.eiar.2004.04.003

[4] Nilsson M, Dalkmann H. Decision-making and strategic environmental assessment. Journal of Environmental Assessment Policy and Management. 2001;**3**:305-327

[5] Therivel R, Partidario MR. The Practice of Strategic Environmental Assessment. London: Earthscan; 1996

[6] Therivel R. Strategic Environmental Assessment. London, UK: Earthscan; 1992

[7] White L, Noble B. Strategic environmental assessment for sustainability: A review of a decade of academic research. Environmental Impact Assessment Review. 2013;**42**:60-66. DOI: 10.1016/j.eiar.2012.10.003

[8] European Strategic Environmental Assessment Directive 2001/42/EC. The European Parliament, Luxembourg; 2011

[9] Arbter K. SEA of Waste Management Plans – An Austrian Case Study. Implementing Strategic Environmental Assessment; 2005. pp. 621-630

[10] Desmond M. Identification and development of waste management alternatives for Strategic Environmental Assessment (SEA). Environmental Impact Assessment Review. 2009;**29**:51-59. DOI: 10.1016/j.eiar.2008.05.003

[11] Josimović B, Marić I, Milijić S. Multi-criteria evaluation in strategic environmental assessment for waste management plan, a case study: The city of Belgrade. Waste Management. 2015;**36**:331-342. DOI: 10.1016/j.wasman.2014.11.003

[12] Josimović B, Marić I. Methodology for the regional landfill site selection. Sustainable Development – Authoritative and Leading Edge Content for Environmental Management. 2012:513-538. DOI: 10.5772/45926

[13] Salhofer S, Wassermann G, Binner E. Strategic environmental assessment as an approach to assess waste management systems. Experiences from an Austrian case study. Environmental Modelling & Software. 2007;**22**:610-618 DOI: 10.1016/j.envsoft.2005.12.031

[14] Bjorklund A, Finnveden G. Life cycle assessment of a national policy proposal – The case of a Swedish waste incineration tax. Waste Management. 2007;**27**:1046-1058. DOI: 10.1016/j.wasman.2007.02.027

[15] Bond R, Curran J, Kirkpatrick C, Lee N, Francis P. Integrated impact assessment for sustainable development: A case study approach. World Development. 2001;**29**:1011-1024. DOI: 10.1016/S0305-750X(01)00023-7

[16] Laurent A, Clavrul J, Bernstd A, Bakas I, Niero M, Gentil E, Christnsen T, Hauschild M. Review of LCA studies of solid waste management systems – Part II: Methodological guidance for a better practice. Waste Management. 2013;**34**:589-606. DOI: 10.1016/j.wasman.2013.12.004

[17] Tukker A. Life cycle assessment as a tool in environmental impact assessment. Environmental Impact Assessment Review. 2000;**20**:435-456. DOI: 10.1016/S0195-9255(99)00045-1

[18] Josimović B, Pucar M. The strategic environmental impact assessment of electric wind energy plants: Case study 'Bavaniste' (Serbia). Renewable Energy. 2010;**35**:1509-1519. DOI: 10.1016/j.renene.2009.12.005

[19] Finnveden G, Nilsson M, Johansson J, Persson A, Moberg A, Carlsson T. Strategic environmental assessment methodologies and applications within the energy sector. Environmental Impact Assessment Review. 2003;**23**:91-123. DOI: 10.1016/S0195-9255(02)00089-6

[20] Energy Sector Development Strategy of the Republic of Serbia by 2025 with projections until 2030 ("Official Gazette of the Republic of Serbia", No. 101/15)

[21] Liou ML, Yeh SC, Yu YH. Reconstruction and systemization of the methodologies for strategic environmental assessment in Taiwan. Environmental Impact Assessment Review. 2005;**26**:170-184. DOI: 10.1016/j.eiar.2005.08.003

[22] Brown AL, Therivel R. Principles to guide the development of strategic environmental assessment. Impact Assessment and Project Appraisal. 2000;**18**:183-189. DOI: 10.3152/147154600781767385

[23] Partidario MR. Elements of an SEA framework – improving the added-value of SEA. Environmental Impact Assessment Review. 2000;**20**:647-663. DOI: 10.1016/S0195-9255(00)00069-X

[24] Therivel R. SEA methodology in practice. In: Therivel R, Partidario MR, editors. The Practice of Strategic Environmental Assessment. London: Earthscan Publications; 1996. pp. 30-44

[25] Partidario MR. Course manual of strategic environmental assessment. The Hague, The Netherlands: IAIA; 2002

[26] Sheate W, Richardson J, Aschemann R, Palerm J, Stehen U. SEA and Integration of the Environment into Strategic Decision-Making. Vol. 1. Main Report to the European Commission, London; 2001

[27] United Nations Environment Program. UNEP'S Environmental Impact Assessment Training Resource Manual. 2nd ed. Geneva, Switzerland/UNEP/ETB Briefs on Economics, Trade and Sustainable Development; 2002

[28] DHV Environment and Infrastructure BV. Existing Strategic Environmental Assessment Methodology. Compiled for the European Commission DGXI, Brussels; 1994

[29] Therivel R. Strategic Environmental Assessment in Action. London: Earthscan Publica tions; 2004

[30] Marsden S. Strategic environmental assessment: An international overview. In: Marsden S, Dovers S, editors. Strategic Environmental Assessment in Australasia. NSW7: The Federation Press; 2002. pp. 1-23

[31] Calvo F, Moreno B, Zamorano M, Szanto M. Environmental diagnosis methodology for municipal waste landfills. Waste Management. 2005;**25**:768-799. DOI: 10.1016/j.wasman.2005.02.019

[32] Higgs G. Integrating multi-criteria techniques with geographical information systems in waste facility location to enhance public participation. Waste Management & Research. 2006;**24**:105-117. DOI: 10.1177/0734242X06063817

[33] Josimović B, Krunić N. Implementation of GIS in selection of locations for regional landfill in the Kolubara Region. SPATIUM. 2008;**17/18**:72-77. DOI: 10.2298/SPAT0818072J

[34] Report on Strategic Environmental Assessment for the Energy Sector Development Strategy of the Republic of Serbia by 2025 with Projections until 2030. Beograd: IAUS; 2013

[35] National Energy Sector Development Strategy ("Official Gazette of the Republic of Serbia", No. 57/08)

[36] Spatial Plan of the Republic of Serbia 2010-2020 (The Official Gazette of the Republic of Serbia, No. 88/2010)

[37] National List of Environmental Indicators ("Official Gazette of the Republic of Serbia", No. 37/11)

[38] Josimović B. Methodological approach to conducting SEA – Initial experiences in Serbia. Planned and Normative Protection of Space and the Environment. Belgrade: Faculty of Geography; 2007

[39] Ananda J, Heralth G. A critical review of multi-criteria decision-making methods with special reference to forest management and planning. Ecological Economics. 2009;**68**:2535-2548. DOI: 10.1016/j.ecolecon.2009.05.010

[40] Figueira J, Greco S, Ehrgott M. In: Salvatore G, editor. Multiple-Criteria Decision Analysis. State of the Art Surveys. Springer; 2005

[41] Kangas J, Kangas A. Multiple criteria decision support in forest management the approach-methods applied, and experiences gained. Forest Ecology and Management. 2005;**207**:133-143. DOI: 10.1016/j.foreco.2004.10.023

[42] Zionts S. MCDM–if not a Roman numeral, then what? Interfaces. 1979;**9**:94-101

[43] Zionts S, Wallenius J. An interactive programming method for solving the multiple criteria problem. Management Science. 1976;**22**:652-663. DOI: 10.1287/mnsc.22.6.652

[44] Simon E. From substantive to procedural rationality. In: Latsis JS, editor. Methods and Appraisal in Economics. Cambridge: Cambridge University Press; 1976

[45] Roy B. Méthodologie multicritere d' aide à la decision. Economica. 1985;**4**:138-140

[46] Banville C, Landry M, Martel JM, Boulaire C. A stakeholder approach to MCDA. Systems Research and Behavioral Science. 1998;**15**:15-32. DOI: 10.1002/(sici)1099-1743(199801/02)15:1<15::aid-sres179>3.0.co;2-b

[47] De Marchi B, Funtowicz O, Lo Cascio S, Munda G. Combining participative and institutional approaches with multicriteria evaluation. An empirical study for water issues in Troina-Sicily. Ecological Economics. 2000;**34**:267-282. DOI: 10.1016/S0921-8009(00)00162-2

[48] Proctor W. MCDA and stakeholder participation – valuing forest resources. In: Getzner M, Spash CL, Stagl S, editors. Alternatives for Environmental Valuation. London: Routledge; 2004. pp. 134-158

[49] Proctor W, Drechsler M. Deliberative multicriteria evaluation. Environment and Planning C: Government and Policy. 2006;**24**:169-190

[50] Munda G. Multiple criteria decision analysis and sustainable development. In: Figueira J, Salvatore G, Ehrgott M, editors. Multiple Criteria Decision Analysis: State of the Art Surveys. New York: Springer; 2005

[51] CLAIRE. Annex 1: The SuRF-UK Indicator Set for Sustainable Remediation Assessment. London; 2011

[52] Linkov I, Satterstrom FK, Kiker G, Batchelor C, Bridges T, Ferguson E. From comparative risk assessment to multi-criteria decision analysis and adaptive management: Recent developments and applications. Environmental Risk Management 2006;**32**:1072-1093. DOI: 10.1016/j.envint.2006.06.013

[53] Rosén L, Söderqvist T, Back PE, Soutukorva Å, Brodd P, Grahn L. Multicriteria analysis (MCA) for sustainable remediation at contaminated sites. Method development and examples. Swedish: Multikriterieanalys (MKA) för hållbar efterbehandling av förorenade områden. Metodutveckling och exempel. Sustainable Remediation Programme, Report 5891Stockholm: Swedish Environmental Protection Agency; 2009

[54] Rosén L, Back PE, Norrman J, Söderqvist T, Brinkhoff P, Volchko Y. Multi-criteria analysis (MCA) for sustainability appraisal of remedial alternatives. In: Proceedings of the Second International Symposium on Bioremediation and Sustainable Environmental Technologies. Jacksonville, Florida, USA; 2013

[55] Cadaster for Small Hydro Power Plants. "Energoprojekt" and the "Jaroslav Černi" Institute for the Development of Water Resources, Belgrade; 1987

[56] Law on the Ratification of the Convention on Environmental Impact Assessment in a Transboundary Context ("Official Gazette of the Republic of Serbia – International Agreements", No. 102/2007)

[57] Protocol on Pollutant Release and Transfer Register—Kiev Protocol. Kiev, Ukraine; 2003

[58] Chatterjee N, Bose G. Selection of vendors for wind farm under fuzzy MCDM environment. International Journal of Industrial Engineering Computations. 2013;**4**:535-546. DOI: 10.5267/j.ijiec.2013.06.002

[59] Khadijah Wan I, Abdullah L. A new Environmental Performance Index using analytic hierarchy process: A case of ASEAN countries. Environmental Skeptics and Critics. 2012;**1**:39-47

Section 5

Renewable Power and Hydrogen Storage Possibilities

Chapter 8

Scalable, Self-Contained Sodium Metal Production Plant for a Hydrogen Fuel Clean Energy Cycle

Alvin G. Stern

Additional information is available at the end of the chapter

http://dx.doi.org/10.5772/67597

Abstract

In this chapter, we present a detailed design study of a novel, scalable, self-contained solar powered electrolytic sodium (Na) metal production plant meant to enable a hydrogen (H_2) fuel, sustainable, closed clean energy cycle. The hydrogen fuel is generated on demand inside a motor vehicle using an efficient hydrogen generation apparatus that safely implements a controlled chemical reaction between either ordinary salinated (sea) or desalinated (fresh) water and sodium metal. The sodium hydroxide (NaOH) byproduct of the hydrogen generating chemical reaction is stored temporarily within the hydrogen generation apparatus and is recovered during motor vehicle refueling to be reprocessed in the self-contained sodium (Na) metal production plant. The electric power for NaOH electrolysis is produced using photovoltaic (PV) device panels spatially arrayed and electrically interconnected on a tower structure that maximizes the use of scarce land area. Our analysis shows that the scalable, self-contained sodium (Na) metal production plant using solar power is technically and economically viable for meeting the hydrogen fuel clean energy needs of all the motor vehicles in the U.S.A. by constructing approximately 450,000 scalable, self-contained sodium (Na) metal production plant units in the southwestern desert region that includes West Texas, New Mexico, Arizona and Southern California.

Keywords: Sodium metal, Sodium metal production plant, Sodium hydroxide, Sodium hydroxide electrolysis, Solar powered electrolysis, Safe hydrogen generation, Hydrogen clean energy cycle

1. Introduction

There is a need in the modern world for sustainable means of producing clean energy economically, on a very large scale. The planet's human population is inexorably increasing toward the 10 billion mark [1–3]. The rapid growth in population presents both opportunities for companies seeking to broaden markets for their products, and challenges for governments, as the growing populations demand their share of prosperity. It is well known that prosperity generating economic growth requires energy [4–6]. To meet the demands for prosperity, carbon based fossil fuel consumption has increased accordingly, resulting in unacceptable levels of air pollution in major conurbations in both advanced and developing countries [7–10]. Much of the pollution comes from burning fossil fuels inside internal combustion engines (ICEs) of motor vehicles and ships. Coal burning thermal power plants used for electricity generation also contribute substantially to the rise in air pollution [11, 12].

Scientific research has hitherto yielded various solutions to the clean energy challenge using innovative approaches ranging from development of hybrid gasoline-electric motor vehicles (HEVs), plug-in hybrid gasoline-electric motor vehicles (PHEVs), pure battery electric vehicles (BEVs), fuel cell electric vehicles (FCEVs), advanced catalysts for reducing exhaust emissions as well as carbon capture technologies applicable to coal burning thermal power plants [13–17]. These solutions however, work only to mitigate the problem of carbon based fossil fuel emissions and do not address the fundamental problem, namely, how to circumvent carbon based fossil fuels in energy generation and ground transport applications. Although HEV, PHEV, BEV and FCEV technologies offer promise in reducing pollution at least locally, they represent at best, an incomplete remedy to a major problem. HEVs and PHEVs still require combustion of gasoline inside an ICE while BEVs require copious quantities of electrical energy for battery charging, generated by power plants connected to the electric grid. The power plants supplying the electric grid can be hydroelectric or nuclear, but far more often, are fossil fuel burning thermal power plants that are only incrementally more efficient than internal combustion engines in motor vehicles [18]. Since hydroelectric generating capacity in the United States of America (U.S.A.) has already been reached and construction of new nuclear power stations is fraught due to well substantiated fears of radiological leaks, it becomes apparent that the only way to meet the increased demand for electricity from electric vehicle proliferation is by constructing more fossil fuel burning thermal power plants [19, 20].

The FCEVs exist at present in small numbers as vehicle prototypes that function primarily as technology demonstrators [21, 22]. FCEVs are unique however, because they represent the only motor vehicle technology that uses hydrogen (H_2) fuel to generate electric energy to power a motor driving the wheels of the vehicle. Although in existence in various forms since the 1960s, FCEVs have not proliferated for manifold reasons, the principal ones being the absence of means for safely storing hydrogen fuel on board, coupled with a lack of means to economically generate sufficiently pure hydrogen ($H_{2(g)}$) fuel to prevent poisoning sensitive catalysts that might be present in the fuel cells [23, 24]. The existing methods of storing hydrogen on board motor vehicles utilize cryogenic storage of liquid hydrogen ($H_{2(l)}$), storage of hydrogen ($H_{2(g)}$) gas at pressures as high as 70 MPa (10,153 psi) in cylinders made from composite material, and storage as a metal hydride (MH_X) in tanks filled with porous metal sponge or powder comprised of light group 1 and 2 metals and/or transition metal elements, namely, Titanium (Ti) or Nickel (Ni) [25–30]. Such direct hydrogen storage methods however, are impractical due to the high cost of suitable transition metals Ti and Ni, and moreover, because an infrastructure is needed to supply hydrogen directly in large volume to fill liquid or gas tanks or to saturate or replenish the metal sponge within the storage reservoir inside a motor vehicle, a procedure fraught with all of the well known safety risks associated with handling large volumes of elemental hydrogen [31, 32]. Furthermore, the existing industrial method of generating hydrogen ($H_{2(g)}$) gas using steam reforming of natural gas, the latter containing mostly methane (CH_4), produces significant quantities of carbon monoxide (CO) even after application of the shift reaction, the latter meant to transform the CO into carbon dioxide (CO_2) [33, 34]. The presence of even minute quantities of CO on the parts per million (ppm) order of magnitude in $H_{2(g)}$ fuel, results in rapid poisoning of sensitive platinum (Pt) catalysts present in the latest generation of low operating temperature, proton exchange membrane (PEM) fuel cells [35]. Catalysts based on a mixture of platinum and ruthenium (Pt-Ru) developed to overcome the sensitivity of pure Pt to carbon monoxide poisoning are not cost effective for large scale application in motor vehicle transport applications due to the dearth of ruthenium [36]. Since hydrogen production by conventional steam reforming methods generates significant quantities of CO and CO_2, it becomes difficult to justify using the approach to generate hydrogen (H_2) fuel for FCEVs given that the purpose of advancing such technology is to eliminate carbon based fossil fuel emissions.

Despite challenges, hydrogen (H_2) which is stored in near limitless quantity in seawater is the only alternative fuel that is more abundant and environmentally cleaner with the potential of having a lower cost than nonrenewable carbon based fossil fuels. We have shown in previous published work that a novel apparatus and method for safely generating hydrogen fuel at the time and point of use from ordinary salinated (sea) or desalinated (fresh) water (H_2O) will enable a vehicle range exceeding 300 miles per fueling using direct combustion of the H_2 fuel in appropriately configured internal combustion engines of the Otto or Diesel types, which is comparable to the vehicle ranges presently achieved with gasoline or Diesel fuels, while providing a sustainable, closed clean energy cycle [37]. The novel hydrogen generation apparatus enables hydrogen ($H_{2(g)}$) fuel to be produced on demand in the motor vehicle using a controlled chemical reaction where liquid water ($H_2O_{(l)}$) is made to react with solid sodium ($Na_{(s)}$) metal reactant to produce hydrogen ($H_{2(g)}$) gas and sodium hydroxide ($NaOH_{(s)}$) byproduct according to Eq. (1).

$$2Na_{(s)} + 2H_2O_{(l)} \rightarrow H_{2(g)} + 2NaOH_{(s)} \tag{1}$$

The high purity hydrogen ($H_{2(g)}$) fuel produced on demand by the novel hydrogen generation apparatus can be used to safely power FCEVs without contaminating the sensitive Pt catalysts present in PEM fuel cells or any other types of catalysts in fuel cells, because the hydrogen is not derived from carbon based fossil fuels, and therefore does not contain even trace amounts of carbon monoxide or sulfur compounds. The seawater reactant can be concentrated to as much as 252.18 grams of sea salt solute per kilogram of seawater solution to provide a fusion temperature T_{Eu} = –21.2 °C (251.95 K), that is equivalent to the eutectic temperature of a 23.18% by weight NaCl in NaCl-H_2O solution [38, 39]. The concentrated sea salt in seawater solution allows the hydrogen generator to operate reliably over a wide ambient temperature range from –21.2 °C (251.95 K) to 56.7 °C (329.85 K) prevailing in the 48 conterminous states of the U.S.A. [37]. The sodium hydroxide (NaOH) byproduct of the hydrogen generating chemical reaction is stored temporarily within the hydrogen generation apparatus and is recovered during motor vehicle refueling. The NaOH is subsequently reprocessed by electrolysis to recover the sodium (Na) metal for reuse in generating hydrogen fuel.

In this chapter, we describe in detail our company's design approach for constructing a novel, scalable, self-contained electrolytic sodium (Na) metal production plant that uses electric power sourced from the sun. The solar powered electrolytic production plant is meant to form an integral part of a hydrogen fuel, sustainable, closed clean energy cycle in conjunction with the novel, hydrogen generation apparatus, enabling Na metal to be produced cost effectively without negative impact to the environment [37].

2. Sodium metal production plant characteristics

For its successful implementation, the hydrogen fuel, sustainable, closed clean energy cycle requires a means of producing quantities of sodium (Na) metal cost effectively on a large scale by electrolysis of sodium hydroxide (NaOH), the latter created as a byproduct of hydrogen ($H_{2(g)}$) fuel generation inside motor vehicles according to Eq. (1). The electrolysis is performed either on pure sodium hydroxide (NaOH) or on a mixture of NaOH and sea salt, the latter consisting primarily of sodium chloride (NaCl), according to Eqs. (2) and (3) [40–42].

$$4Na^+ + 4OH^- \rightarrow 4Na_{(l)} + 2H_2O_{(g)} + O_{2(g)} \tag{2}$$

$$2Na^+ + 2Cl^- \rightarrow 2Na_{(l)} + Cl_{2(g)} \tag{3}$$

The electrical cost of electrolysis can be estimated from the standard reduction potentials of the oxidation and reduction half reactions that occur at the anode and cathode, respectively of the electrolysis cell when implementing Eqs. (2) and (3) [43].

$$\text{Cathode (reduction): } Na^+_{(aq)} + e^- \rightarrow Na_{(s)} \qquad E^\circ_r = -2.71\text{ V} \tag{4}$$

$$\text{Anode (oxidation): } 2Cl^-_{(aq)} \rightarrow Cl_{2(g)} + 2e^- \qquad E^\circ_o = -1.36\text{ V} \tag{5}$$

$$4OH^-_{(aq)} \rightarrow O_{2(g)} + 2H_2O + 4e^- \qquad E^\circ_o = -0.40\text{ V} \tag{6}$$

From Eqs. (4)–(6), the minimum potentials of $E^\circ_{ov} = -4.07$ V and $E^\circ_{ov} = -3.11$ V are needed to electrolyze NaCl and NaOH, respectively. These voltages are significantly higher than the potential of $E^\circ_{ov} = -1.23$ V needed to electrolyze ordinary $H_2O_{(l)}$ to produce $H_{2(g)}$ at the cathode and $O_{2(g)}$ at the anode, however, the benefit from not having to store volatile $H_{2(g)}$ in very large industrial quantities and to transport it between the production plants and refueling stations, outweighs the added electrical cost of producing the solid $Na_{(s)}$ metal.

In the United States of America, the only clean renewable source of energy available in sufficient abundance to implement Eqs. (2) and (3) on a large scale is the radiant energy from the sun that illuminates vast tracts of flat, arid, desert land in West Texas, New Mexico, Arizona and Southern California. The weather in the southwestern U.S.A. is mostly warm and arid with high solar irradiance all year and therefore, constitutes the ideal location for constructing scalable, self-contained solar powered electrolytic sodium (Na) metal production plant units by the thousands [44–48]. Each sodium (Na) metal production plant has to be capable of operating autonomously as a self-contained factory, requiring minimal maintenance and resources. The diagram showing all of the material and energy inputs and outputs of the self-contained sodium (Na) metal production plant is presented in **Figure 1**.

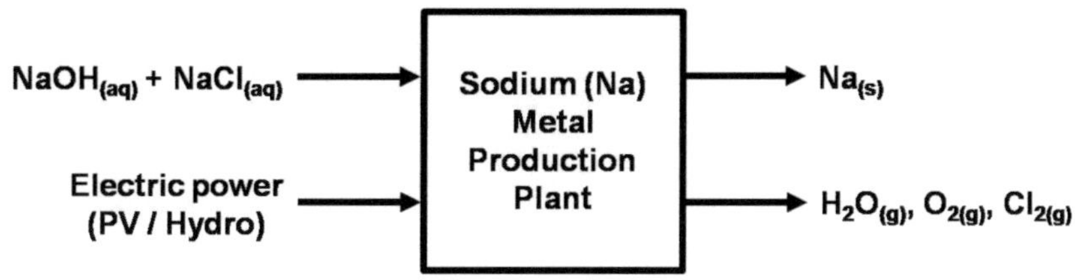

Figure 1. Self-contained sodium (Na) metal production plant operating resources diagram.

In **Figure 1**, electric power for the Na metal production plant is produced using photovoltaic (PV) device panels spatially arrayed and electrically interconnected on a vertical tower structure that maximizes the use of scarce real estate or land area. Up to N_P = 30,000 PV panels, each having an active area $A_P = 1\ m^2$ are mechanically assembled onto the tower, yielding a total PV device panel array active area given as $A_{PA} = N_P \times A_P$ = (30,000 PV panels) × (1 m^2) = 30,000 m^2. Sodium hydroxide (NaOH) or a mixture of NaOH and NaCl recovered from motor vehicle hydrogen generators during refueling, must be supplied to the electrolytic cells to replenish the consumed reactants. When operating the hydrogen generation apparatus in warm tropical climates, NaOH exclusively can be recovered from motor vehicle hydrogen generators during refueling because desalinated (fresh) water ($H_2O_{(l)}$) can safely be used as a reactant without risk for it to freeze. Sodium (Na) metal is produced at the electrolytic cell cathode, and steam ($H_2O_{(g)}$), oxygen ($O_{2(g)}$) gas and some chlorine ($Cl_{2(g)}$) gas are produced at the cell anode, the latter resulting from electrolysis of the NaCl in sea salt. The $H_2O_{(g)}$ and $O_{2(g)}$ can be released directly to the atmosphere while the $Cl_{2(g)}$ must be collected, condensed to a liquid and stored in bottles, for subsequent sale to customers that consume chlorine including the paper and polymer (plastic) manufacturing industries [49]. It is also possible to collect and condense the steam generated at the anode and use the liquid water ($H_2O_{(l)}$) for crop irrigation in arid, desert environments where water resources are limited. The layout of the self-contained sodium (Na) metal production plant is shown in **Figure 2**.

The self-contained sodium (Na) metal production plant shown in **Figure 2**, consists of a solar tower that comprises a photovoltaic (PV) device panel array active area given as A_{PA} = 30,000 m^2. It also consists of a prefabricated Quonset or Q-type metal building having a semicircular cross section, assembled onto a concrete pad foundation that houses electrical switch gear, voltage step down DC-DC converters, the sodium hydroxide (NaOH) electrolytic cells, sodium (Na) metal packaging unit and chlorine (Cl_2) gas separation and bottling unit. A control room permits monitoring the operation of the plant. Two above ground storage tanks are shown located outside of the Q-type metal building with one on either side, for storing aqueous sodium hydroxide ($NaOH_{(aq)}$) solution. Each above ground storage reservoir has a liquid volume capacity of $V_{NaOH(aq)}$ = 37,850 L (10,000 Gal), and is used to replenish the NaOH in the electrolytic cells after the reactant in the cells has been consumed by electrolysis.

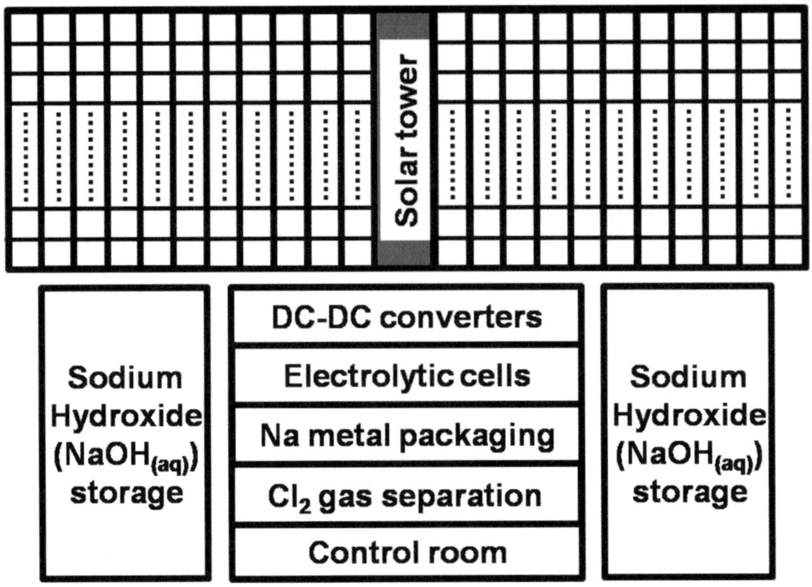

Figure 2. Layout of the self-contained sodium (Na) metal production plant (NOT TO SCALE).

Many factors influence the production yield of sodium (Na) metal during a normal day of plant operation. The two most important factors include the power conversion efficiency of the PV device panels and the magnitude and duration of solar irradiance incident on the PV panels. Other factors affecting the Na metal yield include the power conversion efficiency of the voltage step down DC-DC converter and the efficiency of the electrolytic cell in recovering the Na metal from fused $NaOH_{(l)}$ or from a mixture of fused $NaOH_{(l)}$ and $NaCl_{(l)}$. The efficiency of the voltage step down DC-DC converter depends mainly on how much power is dissipated or lost in the solid state transistors as a result of high frequency on-off switching. The efficiency of the electrolytic cell in this work is viewed in terms of the number of electrons from the electrolytic cell current I_{CELL}, flowing through an electrolytic cell actually needed to produce an atom of Na metal. An alternate definition of the electrolytic cell efficiency might consider the theoretical electric power required to be supplied to the electrolytic cell to produce one mole of Na metal as calculated from the known Gibbs free energy of $NaOH_{(l)}$, divided by the actual power required to be supplied to the electrolytic cell to produce one mole of the Na metal, a definition that we do not consider here. Ideally, each electron flowing through the electric circuit comprising the electrolytic cell should produce one atom of Na metal. In the analysis that follows, it will be assumed that the voltage step down DC-DC converters and the electrolytic cells have efficiencies $\eta_{DC\text{-}DC} = 100\%$ and $\eta_{CELL} = 100\%$, respectively.

2.1. Scalable photovoltaic tower concept

The solar tower comprising the photovoltaic (PV) device panel array supplies electric power to the sodium (Na) metal producing electrolytic cells. The amount of electric energy supplied by the solar tower is a key determinant of the quantity of Na metal that can be electrochemically separated from the NaOH reactant. The energy conversion efficiency of the photovoltaic (PV) device panels is therefore a critical parameter for determining the amount of Na metal that can be produced by the self-contained sodium (Na) metal production plant. Ideally, the photovoltaic (PV) device panels of the solar tower should have maximum optical to electric energy conversion efficiency approaching the thermodynamic limit $\eta_{PVmax} = 93\%$ [50, 51]. According to the Shockley-Queisser theory, such a high conversion efficiency requires PV devices comprising manifold semiconductor junctions [52, 50]. Contemporary, commercially available single junction, monocrystalline silicon photovoltaic (PV) devices attain energy conversion efficiencies ranging between $\eta_{PV} = 15$–18% for front-illuminated silicon devices and up to $\eta_{PV} = 21.5\%$ for back-illuminated silicon devices as summarized in **Table 1**.

	PV panel technology	PV panel area (m^2)	Module efficiency (%)	V_{OC} (V)	V_{MPP} (V)	I_{MPP} (A)	I_{SC} (A)
Sunpower SPR-X21-345	Monocrystalline Si	1.63	21.5	68.2	57.3	6.02	6.39
Suntech STP290S-20	Monocrystalline Si	1.63	17.8	39.8	31.7	9.15	9.55
LG LG280S1C-B3	Monocrystalline Si	1.64	17.1	38.8	31.9	8.78	9.33
First Solar FS-4117-2	Thin film CdTe	0.72	16.3	88.2	71.2	1.65	1.79
Sharp NU-U240F2	Monocrystalline Si	1.63	14.7	37.4	30.1	7.98	8.65

[a]PV panel performance data sourced from respective product datasheets. ASTM AM 1.5G solar irradiance of 1000 W/m^2.

Table 1. Performance of commercial PV device panels in 2016[a].

In **Table 1**, the PV device panels from the Sunpower manufacturing company stand out as having the highest efficiency, due to the use of back-illumination of the silicon device layer. Multijunction devices that offer higher efficiencies remain at a research and development stage and have not reached a level of maturity or cost effectiveness to be ready for release as commercial PV panel products [53]. Our company, AG STERN, LLC is researching the development of advanced, high efficiency PV devices based on novel, very high transmittance, back-illuminated, silicon-on-sapphire semiconductor substrates expected capable of transmitting 93.7% of the total solar irradiance into the semiconductor device layer and therefore capable of achieving an energy conversion efficiency $\eta_{PV} = 90\%$ with proper engineering of the semiconductor photon absorbing layers and PV device structure [54–57].

The solar tower providing power to the electrolytic cells must be scalable in electric power output, and thus capable of allowing PV device panels to be replaced as higher efficiency ones become available, without affecting the overall operation of the self-contained Na metal production plant in any way, other than increasing the Na metal yield. Since the highest performing PV device panels currently offered commercially are listed in **Table 1**, it is possible to calculate the expected power output of the solar tower comprising a PV device panel array active area of $A_{PA} = 30{,}000\ m^2$, under ASTM direct normal air mass (AM) 1.5D standard terrestrial solar spectral irradiance with a total irradiance $Irr_{AM1.5D} = 887\ W/m^2$, as a function of the PV device efficiency, as shown in **Figure 3**.

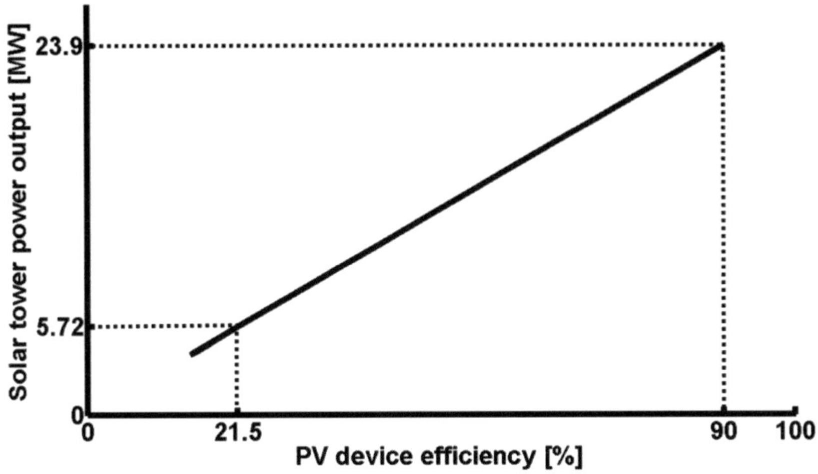

Figure 3. Electric power output for a solar tower PV device panel array with area A_{PA} = 30,000 m^2 under ASTM direct normal AM 1.5D standard terrestrial solar spectral irradiance as a function of the PV device efficiency.

It is clear in **Figure 3**, that single junction, monocrystalline silicon PV devices having an efficiency $\eta_{PV} = 21.5\%$ can be arrayed to generate electric power given as $P_{ST} = 5.72$ MW for electrolysis. Once $\eta_{PV} = 90\%$ efficient PV device panels will become available, the electric power output can be increased to a substantial $P_{ST} = 23.9$ MW, entailing that just 50 of the self-contained sodium (Na) metal production plants can generate peak power given as $P_{ST\text{-}50} = 50 \times 23.9$ MW $= 1195$ MW, matching the power generating capacity of a large commercial nuclear power station.

2.2. Solar irradiance conditions

The electric power output of the solar tower comprising photovoltaic (PV) device panels depends on the magnitude and duration of solar irradiance incident on the PV panels, in addition to the PV device energy conversion efficiency described in Section 2.1. It can be assumed that the PV device energy conversion efficiency is constant and might only decrease in value slowly over time [58]. In contrast, the solar irradiance incident on the PV device panels can vary on a daily basis and is determined by two principal factors, namely, the solar geometry and prevailing atmospheric or meteorological conditions.

The sun is effectively a large hydrogen (H_2) fusion reactor, spherical in shape with a radius $R_{sun} = 6.96 \times 10^8$ m located at a mean distance from the earth given as $r_{sun\text{-}m} = 1.496 \times 10^{11}$ m [59, 60]. The total power output of the sun is given as $P_{sun} = 3.8 \times 10^{26}$ W that radiates isotropically in all directions, resulting in a surface temperature of the solar black body $T_{sun} =$ 5800 K. Of the total solar power output, earth receives only $P_{earth} = 1.7 \times 10^{17}$ W which if converted to electric power, vastly exceeds the earth's energy needs [61]. The sun therefore constitutes an excellent potential source of clean radiant energy to be harnessed on earth.

The solar geometry has a key role in determining how much radiant energy from the sun will be incident on the PV devices located on earth. To understand how the solar geometry influences the solar irradiance at the earth's surface, it will be assumed that earth follows a stable elliptical orbit around the sun described by Eq. (7) [59].

$$r_{sun}(\theta) = \frac{a(1 - e^2)}{1 + e \cdot \cos(\theta)} \tag{7}$$

$$\text{Perihelion: } r_{sun}(0^\circ) = a(1 - e)$$

$$\text{Aphelion: } r_{sun}(180^\circ) = a(1 + e)$$

In Eq. (7), the distance r_{sun}, represents the distance from the center of the sun to the center of the earth and the angle θ, represents the angle between the present position of the earth in orbit around the sun, and the perihelion position when it is closest to the sun. The eccentricity $e = 0.01673$, describes the shape of the elliptical orbit of the earth around the sun and the value a, represents the length of the semi-major axis of the orbit defined as the mean distance from the center of the sun to the center of the earth given as $a = \{r_{sun}(0°) + r_{sun}(180°)\} / 2 = r_{sun\text{-}m} = 1.496 \times 10^{11}$ m. The period of earth's elliptical rotation around the sun is approximately $T_{es} = 365.24$ days and the period of rotation around its own axis is approximately $T_{ea} = 86{,}400$ seconds or 24 hours, the latter known as a *mean solar day* [62–66]. The earth's angular velocity about its own axis is given as $\omega_{ea} = 7.292115 \times 10^{-5}$ rad/sec, although the rotation is slowing over time [67]. The obliquity or tilt of the earth's axis of rotation with respect to a line perpendicular to the plane of its elliptical orbit around the sun is given as $\varepsilon = 23.44°$, although slight precession and nutation of the axis exists [60]. The *plane of the sun* passes through the center of the sun and remains parallel to the earth's *equator* during the elliptical orbit. The *summer and winter solstices* occur approximately on June 21 and December 21, respectively. The *spring and fall equinoxes* occur approximately on March 21 and September 21, respectively, and are characterized by the length of day being equal to the length of night and the earth's equator coinciding with the *plane of the sun*.

The solar declination angle δ, is the angle made by a ray of the sun (passing through the center of the earth and the center of the sun), and the equatorial plane of the earth. The solar declination angle has a range given as $-23.44° \leq \delta \leq +23.44°$. On the *summer solstice* day when $\delta = +23.44°$, the sun shines most directly on the earth's latitude $\varphi_{T\text{-}CAN} = +23.44°$, known as the *Tropic of Cancer*. On the *winter solstice* day when $\delta = -23.44°$, the sun shines most directly on the earth's latitude $\varphi_{T\text{-}CAP} = -23.44°$, known as the *Tropic of Capricorn*. The longitude at the Greenwich prime meridian $\lambda_{PM} = 0°$ [68]. The solar declination angle can be calculated using a geocentric reference frame with the sun located on a *celestial sphere* and earth located at the center of the *celestial sphere*, according to Eq. (8) [60].

$$\delta = arcsin(\sin\lambda_e \cdot \sin\varepsilon) \tag{8}$$

In Eq. (8), λ_e represents the *ecliptic longitude* of the sun, that is the sun's angular position along its apparent orbit in the *plane of the ecliptic*. The *plane of the ecliptic* is tilted by the *obliquity of the ecliptic* angle $\varepsilon = 23.44°$ representing the angle between the *plane of the ecliptic* and the plane of the *celestial equator*, the latter being earth's equator projected onto the *celestial sphere*. The ecliptic intersects the equatorial plane at the points corresponding to the *spring and fall equinoxes* occurring approximately on March 21 and September 21, respectively with $\lambda_e = 0°$ at the March 21 equinox. The Eq. (8) is usually formulated in terms of the day number in a year to account approximately for the elliptic nature of earth's orbit around the sun, as given by Eq. (9) [69].

$$\delta = [0.006918 - 0.399912\cos(\Gamma) + 0.070257\sin(\Gamma) - 0.006758\cos(2\Gamma) + 0.000907\sin(2\Gamma) - 0.002697\cos(3\Gamma) + 0.00148\sin(3\Gamma)](180/\pi) \tag{9}$$

In Eq. (9), the solar declination angle δ, is given as a function of the day angle $\Gamma = 2\pi(N_{day} - 1) / 365$, with day number N_{day}, in a year where $N_{day} = 1$ corresponds to January 1st. The Eq. (9) provides sufficient accuracy due to the relatively small eccentricity value e in Eq. (7). The Eq. (9) shows that the solar illumination of the earth's surface throughout the year occurs most directly between tropical latitudes, namely, the *Tropic of Cancer* ($\delta \approx +23.44°$) and the *Tropic of Capricorn* ($\delta \approx -23.44°$). At the time of the *vernal and autumnal equinoxes*, the earth's equator is located in the *plane of the sun* and every location on earth receives 12 hours of sunlight.

The PV devices can only generate significant electric power during daylight hours therefore, it is essential to verify that a specific geographic location on earth as defined by the latitude φ, and longitude λ, will receive sufficient hours of high incident solar irradiance during the year, before constructing the self-contained sodium (Na) metal production plant. To accurately describe the sun's position in the sky from sunrise to sunset as well as the resulting number of daylight hours, the apparent solar time (AST) is used to express the time of the day. The sun crosses the meridian of the observer at the *local solar noon* when AST = 12, however, the latter time does not coincide exactly with the 12:00 noon time of the locality of the observer. A conversion between the local standard time (LST) and the AST can be made by using equation of time (ET) and longitude corrections. The equation of time accounts for the length of the day variation as a result of the slight eccentricity of the earth's orbit around the sun as well as the tilt of the earth's axis of rotation with respect to a line perpendicular to the plane of its elliptical orbit around the sun. The ET is given by Eq. (10) and has the dimension of minutes [69, 70].

$$\mathrm{ET} = [0.000075 + 0.001868\cos(\Gamma) - 0.032077\sin(\Gamma) - 0.014615\cos(2\Gamma) - 0.04089\sin(2\Gamma)](229.18) \tag{10}$$

The relation between the AST and LST can be made using Eq. (11) that includes the ET and longitude correction which accounts for the fact that the sun traverses 1° of longitude in 4 minutes.

$$\mathrm{AST} = \mathrm{LST} + \mathrm{ET} + (4\ \mathrm{min/deg})(\lambda_S - \lambda_L) - \mathrm{DST} \tag{11}$$

In Eq. (11), the local longitude λ_L, represents the exact longitude value for the location relative to the Greenwich prime meridian of $\lambda_{PM} = 0°$. The standard longitude λ_S, is calculated from λ_L as $\lambda_S = 15° \times (\lambda_L / 15°)$ where the quantity in the parentheses is rounded to an integer value. In Eq. (11), if λ_L is east of λ_S then the longitude correction is positive and if λ_L is west of λ_S then the longitude correction is negative, while DST = 0 minutes or 60 minutes depending on whether daylight savings time applies. If in effect, DST occurs from March until November.

http://dx.doi.org/10.5772/67597

The hour angle h, represents the angular distance in degrees between the longitude λ, of the observer and the longitude whose plane contains the sun, and is calculated according to Eq. (12) [60].

$$h = 15 \cdot (\mathrm{AST} - 12) \tag{12}$$

In Eq. (12), the multiplier of 15° arises because the earth rotates around its own axis 15° in 1 hour, and the AST has a value 0 < AST < 24 hours. When the sun reaches its maximum angle of elevation at the longitude of the observer, it corresponds to the *local solar noon* time where $h = 0°$. The solar elevation angle α_s, above the observer's horizon and its complement, the solar zenith angle θ_{sz}, can be calculated according to Eq. (13) [60].

$$\sin(\alpha_s) = \cos(\theta_{sz}) = \sin(\varphi)\sin(\delta) + \cos(\varphi)\cos(\delta)\cos(h) \tag{13}$$

The solar azimuth angle γ_s, in the horizontal plane of the observer is given by Eq. (14) [60].

$$\cos(\gamma_s) = \frac{\sin(\delta)\cos(\varphi) - \cos(\delta)\sin(\varphi)\cos(h)}{\cos(\alpha_s)} \tag{14}$$

The Eq. (14) uses the convention of the solar azimuth angle defined as positive clockwise from north meaning that east corresponds to 90°, south corresponds to 180° and west corresponds to 270°. The solar azimuth angle provided by Eq. (14) should be interpreted as $0° \leq \gamma_s \leq 180°$ when $h < 0°$, meaning the sun is located east of the observer, and interpreted as $180° \leq \gamma_s \leq 360°$ when $h > 0°$, meaning the sun is located west of the observer. The sunrise equation allows to calculate the number of hours of sunlight per day that depends on the solar declination angle δ, and on latitude φ, as shown in Eq. (15), which is derived from Eq. (13) by setting $\alpha_s = 0°$ and solving for h.

$$H_{ss} = -H_{sr} = \frac{1}{15}\arccos[-\tan(\varphi)\tan(\delta)] \tag{15}$$

In Eq. (15), the hour angle h, has a value $-180° < h < 0°$ at sunrise and $0° < h < +180°$ at sunset, and when divided by 15°/hour, yields the number of hours before and after the *local solar noon* time (AST = 12) when $h = 0°$, corresponding to sunrise H_{sr}, and sunset H_{ss}, respectively. Using Eqs. (9) and (15), it is possible to calculate for any geographic location on earth the number of daylight hours on any specific day of the year. The maximum angle of solar elevation $\alpha_{s\text{-}max}$, above the observer's horizon that occurs at the *local solar noon* time is calculated according to Eqs. (16) and (17) that are derived from Eq. (13) by setting $h = 0°$.

$$\alpha_{s-\max} = 90^{\circ} - (\varphi - \delta) \tag{16}$$

$$\alpha_{s-\max} = 90^{\circ} + (\varphi - \delta) \tag{17}$$

The Eq. (16) is applicable for the northern hemisphere while Eq. (17) is applicable for the southern hemisphere. Formulations for the solar declination, elevation, azimuth and zenith angles which account with greater accuracy for the elliptic nature of earth's orbit around the sun exist in the scientific literature often as part of solar position algorithms however, they can be rather complex

[71–76]. The air mass can be calculated as a function of the true solar zenith angle θ_{sz}, given by Eq. (13), considering the effect of atmospheric refraction, according to Eq. (18) [77].

$$\mathrm{AM} = \frac{1.002432\cos^2(\theta_{sz}) + 0.148386\cos(\theta_{sz}) + 0.0096467}{\cos^3(\theta_{sz}) + 0.149864\cos^2(\theta_{sz}) + 0.0102963\cos(\theta_{sz}) + 0.000303978} \tag{18}$$

The direct normal total (spectrally integrated) solar irradiance as a function of the air mass (AM) and atmospheric conditions that include the effects of elevation, can be calculated using the Parameterization Model C developed by Iqbal, given in Eq. (19) [78–80].

$$Irr_n = 0.9751 \cdot E_0 \cdot Irr_0 \cdot \tau_r \tau_o \tau_g \tau_w \tau_a \tag{19}$$

In Eq. (19), $Irr_0 = 1367\ \mathrm{W/m^2}$, represents the solar constant or total irradiance of the AM 0 standard solar spectral irradiance in space prior to attenuation by earth's atmosphere and the factor 0.9751 is included because the spectral interval of 0.3–3.0 μm is used by the detailed SOLTRAN model from which the Parameterization Model C is derived. The factor E_0, represents the effect of the eccentricity of earth's orbit around the sun on the solar constant Irr_0, due to the periodically varying distance between the earth and sun, as given by Eq. (20) [69].

$$\begin{aligned} E_0 = (r_{\mathrm{sun}-m}/r_{\mathrm{sun}})^2 &= 1.000110 + 0.034221\cos(\Gamma) + 0.001280\sin(\Gamma) \\ &+ 0.000719\cos(2\Gamma) + 0.000077\sin(2\Gamma) \end{aligned} \tag{20}$$

In Eq. (19), the symbol τ_r represents transmittance by Rayleigh scattering, τ_o represents transmittance by ozone, τ_g represents transmittance by uniformly mixed gases, τ_w represents transmittance by water vapor and τ_a represents transmittance by aerosols. The expressions for τ_r, τ_o, τ_g, τ_w, τ_a are given in Eqs. (21–25).

$$\tau_r = \exp[-0.0903 \cdot \mathrm{AM}_a^{0.84}(1.0 + \mathrm{AM}_a - \mathrm{AM}_a^{1.01})] \tag{21}$$

$$\tau_o = 1 - [0.1611 \cdot U_3(1.0 + 139.48 \cdot U_3)^{-0.3035} - 0.002715 \cdot U_3(1.0 + 0.044U_3 + 0.0003U_3^2)^{-1}] \tag{22}$$

$$\tau_g = \exp[-0.0127 \cdot \mathrm{AM}_a^{0.26}] \tag{23}$$

$$\tau_w = 1 - 2.4959 \cdot U_1[(1.0 + 79.034U_1)^{0.6828} + 6.385U_1]^{-1} \tag{24}$$

$$\tau_a = \exp[-k_a^{0.873}(1.0 + k_a - k_a^{0.7088})\mathrm{AM}_a^{0.9108}] \tag{25}$$

In Eqs. (21), (23) and (25), AM_a represents the air mass at the actual atmospheric pressure P and is given as $\mathrm{AM}_a = \mathrm{AM} \times (P / P_0)$ where AM, calculated using Eq. (18), corresponds to standard atmospheric pressure $P_0 = 101325$ Pa. The actual atmospheric pressure P, can be calculated using the isothermal atmosphere formula in Eq. (26).

$$P = P_0 \cdot exp\left[\frac{-g_0 \cdot M_{\mathrm{air}} \cdot h_{\mathrm{PV}}}{R_g T}\right] \tag{26}$$

The Eq. (26) can be used to calculate the atmospheric pressure P in pascals (Pa) at an elevation h_{PV}, in meters (m) above mean sea level, assuming a constant ambient atmospheric temperature T in Kelvin (K), over the difference in elevations. The molar mass of air $M_{air} = 0.028964$ kg/mol, earth's gravitational acceleration near the surface $g_0 = 9.80665$ m/sec^2, and the universal gas constant $R_g = 8.3144621$ J/K·mol [43, 81, 82]. In Eq. (22), $U_3 = l_o \times$ AM, where l_o represents the vertical ozone layer thickness in centimeters (cm) at normal temperature and surface pressure (NTP), and AM is given by Eq. (18). The vertical ozone layer thickness is assumed in the present work to have a mean annual value $l_o = 0.35$ cm (NTP). In Eq. (24), $U_1 = w'(P/101325)^{0.75}\,(273/T)^{0.5} \times \text{AM}$, where w' represents the precipitable water vapor thickness in centimeters (cm) under the actual atmospheric conditions with pressure P in pascals (Pa), and temperature T in Kelvin (K). In Eq. (25), k_a represents the aerosol optical thickness and has a range $0 < k_a < 1$, depending on how clear or aerosol saturated the atmosphere is. The value of k_a is generally measured at the optical wavelengths of 380 nm and 500 nm due to low ozone absorption and calculated using the formula, $k_a = 0.2758 \cdot k_{a|\lambda=380\text{ nm}} + 0.35 \cdot k_{a|\lambda=500\text{ nm}}$. The aerosol optical thickness can vary daily, however, it is assumed in the present work to have a mean annual value $k_a = 0.1$ reflecting clear days.

In **Table 2**, geographic and climate characteristics are specified for four prospective locations of the scalable, self-contained solar powered electrolytic sodium (Na) metal production plant including El Paso, Texas; Alice Springs, Australia; Bangkok, Thailand and Mbandaka, Democratic Republic of Congo (DRC).

	[a]Geographic coordinates φ, λ & elevation h_{PV} (m)	[b]Mean monthly temperature T (°C)	[c]Mean annual precipitable water w' (cm)	Mean annual *local solar noon* air mass (AM)
El Paso	+31.807°, –106.377°/ 1206	17.7	1.27	1.27
Alice Springs	–23.807°, +133.902°/ 545.2	20.5	1.9	1.16
Bangkok	+13.693°, +100.75°/ 1.524	28.2	4.52	1.08
Mbandaka	+0.0225°, +18.288°/ 316.9	25.1	3.81	1.04

[a]Accurate geographic coordinates and elevations above mean sea level for largest commercial airports of respective cities.
[b][83].
[c][84 87].

Table 2. Geographic and climate characteristics for four prospective plant locations.

In **Table 2**, the geographic coordinates use the sign convention of +/– latitude φ, for locations north and south of the equator, respectively, and +/– longitude λ, for locations east and west of the prime meridian ($\lambda_{PM} = 0°$). According to the Köppen-Geiger climate classification system, El Paso has an arid, desert, cold (BWk) climate, Alice Springs has an arid, desert, hot (BWh) climate, Bangkok has a tropical, savanna (Aw) climate and Mbandaka has a tropical, rainforest (Af) climate [88]. The mean annual *local solar noon* air mass values in **Table 2** are calculated using Eqs. (9) and (16)–(18), and the length of day results provided by Eq. (15), show that proximity to the equator for the self-contained sodium (Na) metal production plant maximizes the solar irradiance and provides uniform hours of daylight per day throughout the year. Tropical regions however, experience higher mean monthly temperatures and consequently higher mean annual precipitable water vapor levels with a long rainy season and sun obscuring rain clouds. Other factors that can significantly affect incident solar irradiance on the PV device panels include the geographic elevation above mean sea level and aerosols comprising solid and liquid sunlight obscuring particulates in the air that can occur naturally due to dust storms and volcanic activity or from human activity such as slash and burn agriculture [89]. According to the accumulated world meteorological data, the southwestern region of the U.S.A. and the central region of Australia, both constitute nearly ideal locations for the self-contained sodium (Na) metal production plants due to the existence of an arid, desert climate, vast tracts of flat open land, sparse human population and many clear days with high solar irradiance throughout the year [47, 48, 90, 91].

3. Sodium metal production plant architecture

The architecture of the self-contained sodium (Na) metal production plant has to provide an optimal balance between high performance, reliability and cost effective operation. The highest performance can be achieved by constructing the solar tower of the plant shown in **Figure 2**, in a manner that allows the optical k-vectors from the sun to be normally incident onto the photovoltaic (PV) device panels throughout the entire period of daylight from sunrise to sunset. The **Figure 4** shows a solar tower architecture with fixed PV device panels, wherein the panels can rotate and tilt with the solar tower as a single unit to follow the sun's overhead trajectory.

In **Figure 4**, the solar tower is fabricated using modular sections comprised of high strength, lightweight aluminum alloy that can be fitted end to end and bolted together until the final slant height of the structure $S_h = 300$ m is achieved. Up to $N_{B\text{-}L/R} = 100$ branches that constitute levels, comprised of modular sections each having a final length $B_l = 50$ m, extend out on each side from the central column of the solar tower resulting in a width for the structure of approximately $S_w = 100$ m. The PV device panels are mounted along the top and bottom of the branches projecting from the central column of the solar tower. The projecting branches of the solar tower also enable ground glass, light diffusing panels to be placed between the rows of solar panels that allow sunlight to penetrate and effectively illuminate the land area beneath the solar tower to support crop cultivation including rice paddy fields. The central column of the solar tower is fixed at the base to a static pylon and approximately midway up the height of the column, to a boom for elevating the solar tower to a tower elevation angle $\pi/2 - \alpha_s$, that is the complement of the solar elevation angle α_s, the latter given in Eq. (13). The opposite end of the boom is fixed to the center of a railroad flatcar mounted on a linear train track that is capable of being displaced along the track to elevate and lower the solar tower. An electric motor module mounted midway up the height of the column enables the PV device panels to collectively rotate left and right about the axis of the central column to track the solar azimuth angle γ_s, the latter given in Eq. (14), thereby allowing the optical k-vectors from the sun to impinge at normal incidence onto the PV device panels throughout most of the day of operation. When the sun has set at the end of the day or when inclement weather of sufficient severity is expected, the solar tower can be lowered to be parallel and nearly flush with the ground as shown in **Figure 4**, with metal louvers drawn over the PV device panels to safeguard against damage to the panels from storms, strong winds and flying debris that can occur, albeit rarely, in the southwestern U.S.A. [92, 93].

http://dx.doi.org/10.5772/67597

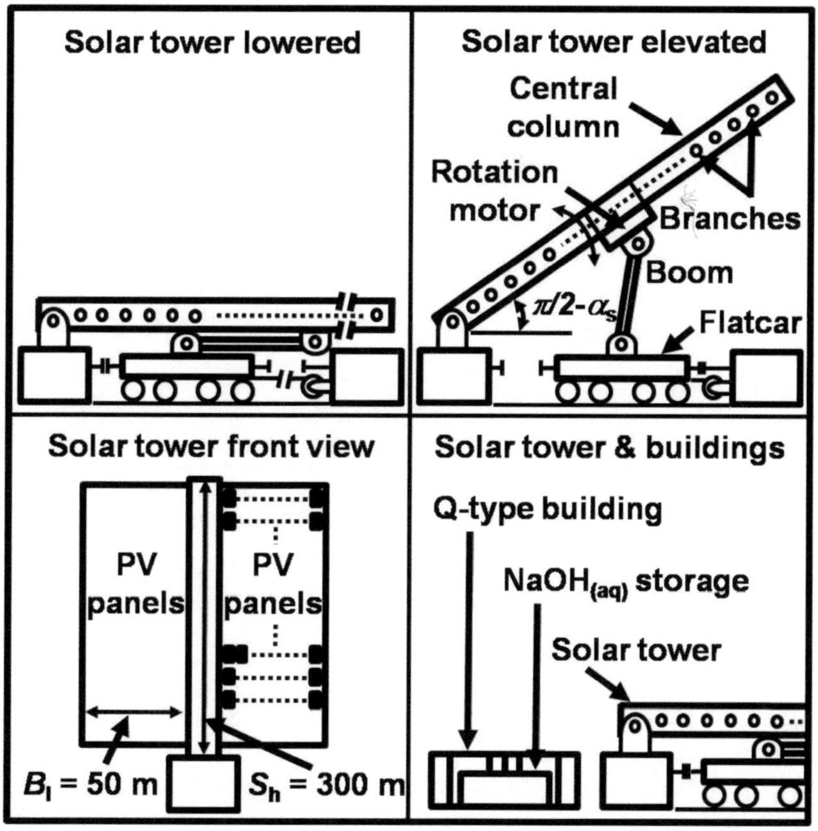

Figure 4. Solar tower architecture (NOT TO SCALE).

3.1. Electrical design of the solar tower

The electrical design of the solar tower comprising PV device panels has to accommodate scalability in the power output level, where it is possible to supplant the existing PV device panels with newer and more efficient ones when they become available, without having to modify other components in the solar tower. It is therefore necessary to optimally dimension the electrical conductors embedded within the branches and central column that transmit the electric power generated by the photovoltaic (PV) device panels to the electrolytic cells, according to the magnitude of the current expected to be transmitted once $\eta_{PV} = 90\%$ efficient PV device panels become available to be installed on the branches of the tower as shown in **Figure 4**. To develop an accurate design for the aluminum current carrying conductors of the solar tower, it becomes necessary to define the electrical interconnection topology of the photovoltaic (PV) device panels installed on the solar tower. In **Figure 5**, the equivalent circuit model of the photovoltaic (PV) device panel array installed on the solar tower is shown, with the corresponding straight line approximation of the current versus voltage curve for the PV device panel array.

In the circuit models shown in **Figure 5**, I_{PV} represents the PV device current, V_{OC} represents the PV device open circuit voltage, R_P represents the parallel resistance of the PV device that ideally

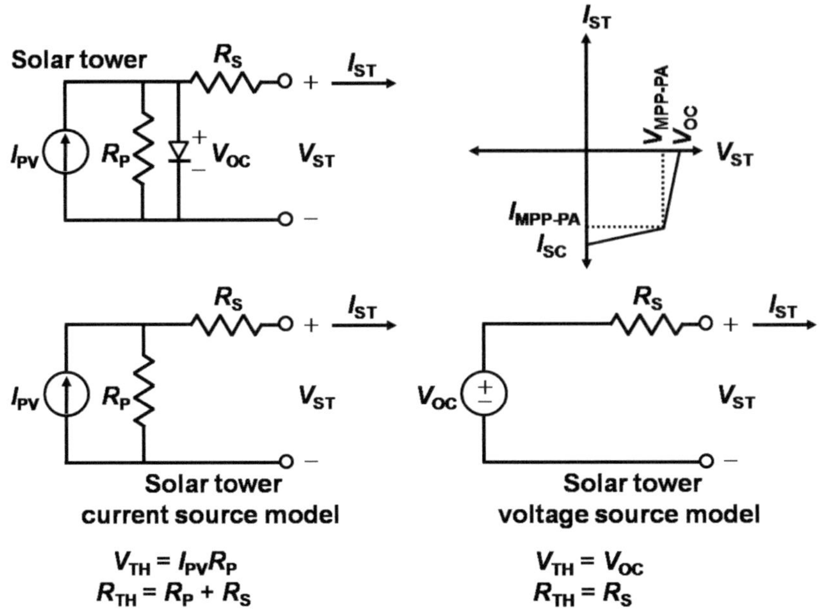

Figure 5. Circuit model of the photovoltaic (PV) device panel array installed on the solar tower.

should be infinite, R_S represents the series resistance of the PV device that ideally should be zero in value, I_{ST} represents the output current of the solar tower and V_{ST} represents the output voltage of the solar tower. The circuit models shown in **Figure 5** are applicable for a single PV cell as well as for an array of PV device panels comprised of PV cells connected in series and/or in parallel [94]. The expressions for the Thevenin equivalent circuit voltage V_{TH} and resistance R_{TH}, are given both for the linear equivalent current source and voltage source circuits. The electrical interconnection topology of the PV device panels on the solar tower must achieve an optimal balance in operating parameters including the solar tower supply system maximum voltage $V_{ST\text{-}MAX}$ and maximum current $I_{ST\text{-}MAX}$. It will be assumed in the further analysis and calculations that $V_{ST\text{-}MAX}$ corresponds to the maximum power point (MPP) operating voltage of the PV device panel array ($V_{MPP\text{-}PA}$) rather than to the open circuit voltage V_{OC}, of the array and similarly, that $I_{ST\text{-}MAX}$ corresponds to the MPP operating current of the PV device panel array ($I_{MPP\text{-}PA}$) rather than to the short circuit current I_{SC}, of the array. The assumptions can be considered valid since it is possible to show using Eqs. (27)–(29) that there exists only a small difference in the value between $V_{MPP\text{-}PA}$ and V_{OC} and similarly between $I_{MPP\text{-}PA}$ and I_{SC} if R_P is large and R_S is small. The Eqs. (27)–(29) describe the solar tower PV array model shown in **Figure 5**.

$$I_{PV} = I_{SC}\frac{(R_S + R_P)}{R_P} \tag{27}$$

$$\frac{V_{MPP-PA}}{I_{SC} - I_{MPP-PA}} = R_S + R_P \tag{28}$$

$$\frac{V_{OC} - V_{MPP-PA}}{I_{MPP-PA}} = R_S \tag{29}$$

The Eq. (27), can be derived from the circuit model in **Figure 5** by short circuiting the output terminals of the solar tower where $I_{ST} = I_{SC}$. The Eq. (28) is calculated from the solar tower PV array operating as a current source at the maximum power point (MPP), and Eq. (29) is calculated from the solar tower PV array operating as a voltage source at the MPP.

The solar tower must be capable of transmitting $P_{ST} = 23.9$ MW of electric power as indicated in **Figure 3**, from the photovoltaic (PV) device panels to the electrolytic cells of the scalable, sodium (Na) metal production plant while maintaining reasonable design values for $V_{ST\text{-}MAX}$ and $I_{ST\text{-}MAX}$ that will not escalate the cost of the system. To achieve an optimal balance between the $V_{ST\text{-}MAX}$ and $I_{ST\text{-}MAX}$ parameters of the solar tower supply system, it is necessary to identify the most reliable and cost effective approach for maintaining the PV devices at or near their maximum power point (MPP) during operation. Many methods exist for providing a variable or dynamic load to PV devices and performing maximum power point tracking (MPPT), however, for the present application it is necessary to consider that since there are up to $N_P =$ 30,000 PV device panels installed on the solar tower, it becomes uneconomical to use any approach that requires dedicated MPPT hardware for so many panels individually or even for small groups of panels and therefore, a collective solution is required for the entire PV device panel array having an area $A_{PA} = 30{,}000$ m^2. A reliable method of operating PV devices very near, if not exactly at the MPP can be implemented by controlling the output voltage of the PV devices [95]. The maximum power point output voltage of a single PV cell $V_{MPP\text{-}C}$, is always very near in value to a temperature dependent operating voltage of the PV cell and this phenomenon can be used to implement a type of MPPT by controlling the output voltage V_{ST}, of the entire PV device panel array with area $A_{PA} = 30{,}000$ m^2. The output voltage V_{ST} of the entire PV device panel array can be controlled for example, by using a voltage step down pulse width modulated (PWM) DC-DC converter with closed loop feedback control that is operated at the proper duty cycle required to maintain the input voltage V_{IN}, to the converter (which is equivalent to the output voltage V_{ST}, of the PV device panel array), at the value near to the MPP voltage. In practice, the MPP output voltage of a single junction, monocrystalline silicon PV cell is given as $V_{MPP\text{-}C} \approx 0.4$ V which is too low for direct input to a PWM DC-DC converter [95]. Commercial single junction, front-illuminated, monocrystalline silicon PV panels of the type shown in **Table 1**, typically contain 60 PV cell modules connected in series resulting in a maximum power point output voltage for the panel given as $V_{MPP\text{-}P} = 30.1$ V for the NU-U240F2 Sharp panel, $V_{MPP\text{-}P} = 31.9$ V for the LG280S1C-B3 LG panel and $V_{MPP\text{-}P} = 31.7$ V for the STP290S-20 Suntech panel. The maximum power point output voltage of the 96 module, back-illuminated SPR-X21-345 Sunpower panel is given as $V_{MPP\text{-}P} = 57.3$ V. The series connected PV cell modules use bypass diodes to prevent power loss from an entire chain of series connected cells if one cell becomes shaded, and receives less illumination than other cells causing its resistance to increase, by allowing energy to be collected from PV cells that are not shaded and are outside of the bypassed section of the chain containing the shaded cell(s) [96, 97]. If 15 of the standard 60 cell module single junction, front-illuminated, monocrystalline silicon PV panel types are connected together in series, then the solar tower supply system maximum voltage can be limited to approximately $V_{ST\text{-}MAX} \approx 450$ VDC, which corresponds to a *low voltage* supply according to the International Electrotechnical Commission (IEC) that defines *low voltage* DC equipment as having a nominal voltage below 750 VDC [98]. A *low* *voltage* DC supply system entails a minor risk of electric arcing through the air and an exemption from specialized protection equipment needed for *high voltage*. If 15 of the 96 cell module, back-illuminated SPR-X21-345 Sunpower PV panels are connected together in series, then the solar tower supply system maximum voltage $V_{ST\text{-}MAX} \approx 900$ VDC, which exceeds by 150 V the nominal 750 VDC IEC standard of a *low voltage* supply. Once $\eta_{PV} = 90\%$ efficient PV device panels will become available, it is expected that they will each have an area $A_P = 1\ m^2$ and the number of PV cell modules connected together in series will yield a maximum power point voltage for the panel given as $V_{MPP\text{-}P} \approx 30$ V that is similar in value to most of the commercial single junction, front-illuminated, monocrystalline silicon PV panels listed in **Table 1**, that contain 60 PV cell modules connected in series. Under an ASTM AM 1.5G standard terrestrial solar spectral irradiance with a total irradiance $Irr_{AM1.5G} = 1000\ W/m^2$, the corresponding maximum power point current of the $\eta_{PV} = 90\%$ efficient PV panel can be calculated as $I_{MPP\text{-}P} = (\eta_{PV} \times Irr_{AM1.5G}) / V_{MPP\text{-}P} = 30.0$ A, which is a substantially larger current than the $I_{MPP\text{-}P}$ current values listed in **Table 1**, as might be expected for the more efficient PV panel unit. Therefore, it is necessary to design the current conductors within the scalable solar tower to be capable of transmitting the substantially larger current of more efficient PV panels once they will become available for installation onto the solar tower. The solar tower supply system maximum voltage can be set to $V_{ST\text{-}MAX} = 450$ V, achieved by connecting in series 15 PV device panels with an efficiency $\eta_{PV} = 90\%$ and $V_{MPP\text{-}P} \approx 30$ V, mounted on a single branch section of the solar tower apparatus as shown in **Figure 6**.

In **Figure 6**, each branch section supports the installation of 2 groups of 15 series connected PV panels having an efficiency $\eta_{PV} = 90\%$ for a total $N_{P\text{-}Bsec} = 30$ PV panels per branch section. A total of 5 branch sections should be fastened together end to end, to yield $N_{P\text{-}B} = 5 \times N_{P\text{-}Bsec} = 150$ PV panels per branch with 5 groups of 15 series connected PV panels installed along the top of the branch and another 5 groups of 15 series connected PV panels installed along the bottom of the branch. Each branch section has a length $Bsec_l = 10$ m, and contains embedded within the aluminum current conductors, yielding a branch length $B_l = 50$ m as shown in **Figure 4**. The groups of 15 series connected PV panels are electrically connected to a pair of aluminum conductors inside the branch, resulting in parallel electrical interconnection for the groups of 15 series connected PV panels that increases the electric current delivered by the solar tower. Each PV panel has a width $P_w = 0.66$ m and a height $P_l = 1.5$ m for a total PV panel area given as $A_P = P_w \times P_l = 1\ m^2$. Since the height of each PV panel is given as $P_l = 1.5$ m, then each branch has an overall height given as $B_h = 2 \times P_l = 3$ m. The solar tower that has a height $S_h = 300$ m as shown in **Figure 4**, can therefore accommodate up to 100 branches extending out from the left and right sides of the central column, yielding a total PV panel array area on the solar tower given as $A_{PA} = 2 \times (S_h / B_h) \times N_{P\text{-}B} \times A_P = 30{,}000\ m^2$.

The electric current from each branch section flows into the electrical conductors installed inside the central column of the solar tower as shown in **Figure 6**. Therefore, electric current that flows from the outermost branch section toward the central column increases as each branch section contributes additional current generated by the 30 PV panels mounted on it. If an ASTM direct normal AM 1.5D standard terrestrial solar spectral irradiance with a total irradiance $Irr_{AM1.5D} = 887\ W/m^2$, is incident on the PV panels having an efficiency $\eta_{PV} = 90\%$ and $V_{MPP\text{-}P} = 30$ V, then the maximum power point output current can be calculated as $I_{MPP\text{-}P} = (\eta_{PV} \times Irr_{AM1.5D})$ /

http://dx.doi.org/10.5772/67597

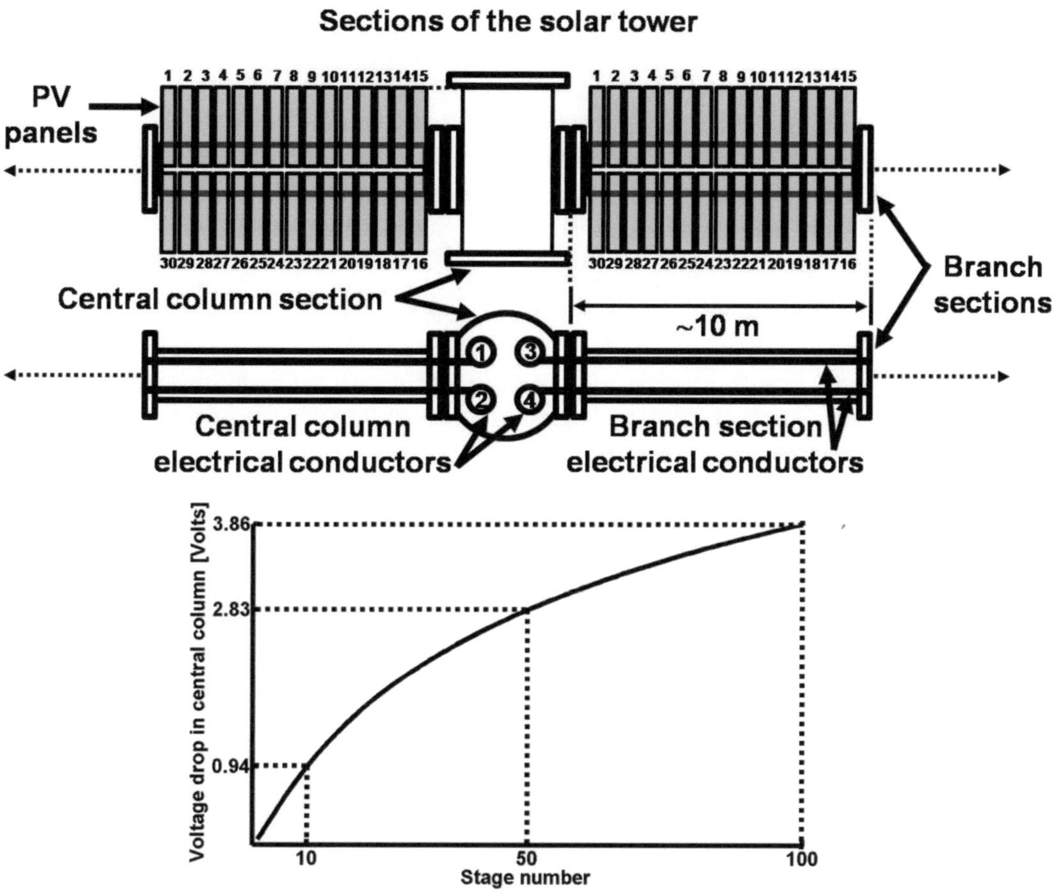

Figure 6. Solar tower branch sections showing groups of 15 series interconnected PV device panels installed.

($V_{MPP\text{-}P}$) = 26.6 A. The result for $I_{MPP\text{-}P}$ = 26.6 A, entails that each branch section will contribute $I_{Bsec} = 2 \times I_{MPP\text{-}P}$ = 53.2 A. Therefore, the current in each branch increases by I_{Bsec} = 53.2 A as it flows in from the outermost branch section toward the current conductors of the central column of the solar tower. The total current contributed by each branch of the solar tower is then calculated as $I_B = 5 \times I_{Bsec}$ = 266 A, since a single branch comprises 5 branch sections. Each side of the solar tower has $N_{B\text{-}L/R}$ = 100 branches for a total number of branches on the solar tower given as $N_{B\text{-}ST} = 2 \times N_{B\text{-}L/R}$ = 200 branches. It is convenient however, to install four electrical conductors labeled 1, 2, 3 and 4, in the central column of the solar tower as shown in **Figure 6**. The first pair of current conductors 1 and 2, transmits current from the branches mounted on the left side of the solar tower and the second pair of current conductors 3 and 4, transmits current from the branches mounted on the right side of the solar tower, so that each conductor pair only needs to transmit a maximum current $I_{ST\text{-}MAX}$ = 26,600 A.

The four electrical conductors installed within the modular sections of the central column of the solar tower that are located at or near the top of the solar tower, do not have to carry the maximum current $I_{ST\text{-}MAX}$ = 26,600 A, rather only the four electrical conductors installed in the bottom most modular section of the central column have to be dimensioned to transmit the maximum current. The diameters of aluminum electrical conductors within the central column sections scale linearly from a diameter $D_{S1} = 2$ cm in stage 1 at the top to $D_{S100} = 41.6$ cm for stage 100 at the bottom of the solar tower, the latter that transmits the maximum current $I_{ST\text{-}MAX} = 26{,}600$ A. **Figure 6**, shows the calculated voltage drop across the aluminum electrical conductors in all 100 modular sections of the central column of the solar tower to be given as $V_{ST\text{-}DROP} = 3.86$ V, corresponding to a low loss considering that the solar tower supply system maximum voltage $V_{ST\text{-}MAX} = 450$ V.

The electrical design of the solar tower described provides manifold advantages including a mostly parallel electrical interconnection architecture for the PV device panels that allows the MPP of the PV device panels to be controlled collectively by controlling the output voltage V_{ST} of the PV device panel array using a voltage step down PWM DC-DC converter designed to have a constant output voltage V_{OUT}, and controllable input voltage V_{IN}, the latter supplied from the PV device panel array and equal to V_{ST}, shown in **Figure 5**. Other advantages include a *low voltage* DC solar tower supply system with optimal balance between the maximum voltage $V_{ST\text{-}MAX} \approx 450$ V and maximum current $I_{ST\text{-}MAX} = 26{,}600$ A, wherein two independent and electrically isolated power supply feeds are provided from the left side branches and right side branches of the solar tower apparatus, respectively. Yet another important advantage of the design includes a low center of gravity for the solar tower apparatus due to the modular sections comprising the central column being heavier near the base and lighter near the top as a result of larger diameter aluminum electrical conductors placed near the base of the tower.

3.2. Electrical design of the sodium hydroxide electrolysis plant

The solar tower apparatus described in Section 3.1 implements a mostly parallel electrical interconnection architecture for the PV device panels that yields two identical, independent and electrically isolated low voltage DC power supplies, each having a maximum voltage $V_{ST\text{-}MAX} \approx 450$ V and maximum current $I_{ST\text{-}MAX} = 26{,}600$ A, wherein the two power supply feeds emanate from the left side branches and right side branches of the solar tower apparatus, respectively as shown in **Figure 6**. The sodium hydroxide (NaOH) electrolysis cell however, requires a substantially lower voltage for operation. The quantity of sodium (Na) metal produced by the NaOH electrolytic cell depends fundamentally on the magnitude of the DC current flowing between the anode and cathode terminals of the electrolysis cell. The Eqs. (4)–(6) provide the standard reduction potentials of the oxidation and reduction half reactions that occur at the anode and cathode, respectively of the electrolytic cell. In practice, the electrolytic cell operating voltage has to be set at approximately $V_{CELL} = 4$ V for electrolysis of pure NaOH or $V_{CELL} = 5$ V for electrolysis of a mixture of NaOH and NaCl, the latter derived from sea salt. The higher voltage accounts for the overvoltage effects [99]. The maximum current that can be supplied to an electrolytic cell operating at approximately $V_{CELL} \approx 5$ V will be $I_{CELL} \approx 100{,}000$ A and is limited in large measure by the material cost as well as by the mass and physical dimensions of the electrical conductors required to transmit such magnitude of the current.

In a fused or molten state, $NaOH_{(l)}$ is highly corrosive and therefore the only conventional materials capable of withstanding prolonged exposure to its caustic effects at an elevated temperature include graphite, iron and nickel. Graphite however, cannot be used as an anode electrode because it will react with the oxygen (O_2) generated to produce carbon dioxide (CO_2) and become consumed in the process. Iron can withstand corrosion from fused $NaOH_{(l)}$, and consequently could be used to fabricate the electrolytic vessel however, as an anode electrode, the reaction with steam ($H_2O_{(g)}$) and O_2 will quickly oxidize and erode iron. The only material suitable for fabricating both the anode and cathode electrodes remains nickel (Ni) which is significantly more expensive than both graphite and iron. The cost of the Ni electrodes therefore becomes an important factor in limiting the maximum current in the electrolytic cell. The other factor limiting the current in the electrolytic cell becomes the PWM DC-DC converter that must supply the large currents at the correct output voltage to the electrolytic cell, safely and reliably.

It is necessary to provide two identical voltage step down PWM DC-DC converters to convert the power $P_{ST\text{-}L/R} = 11.95$ MW delivered by the two independent low voltage DC power supplies of the solar tower, each having a maximum voltage $V_{ST\text{-}MAX} \approx 450$ V and maximum current $I_{ST\text{-}MAX} = 26{,}600$ A. The PWM DC-DC converter must be designed to safely convert the electric power supplied from the solar tower to magnitudes of DC voltage and DC current that are appropriate for supplying the NaOH electrolytic cells. The PWM DC-DC converters and electrolytic cells are housed together inside the prefabricated Q-type metal building indicated in **Figure 4**. If each PWM DC-DC converter is designed to supply a single NaOH electrolytic cell with a power conversion efficiency $\eta_{DC\text{-}DC} = 100\%$, then the maximum current in the single NaOH electrolytic cell would be calculated as $I_{OUT} = (V_{ST\text{-}MAX} \times I_{ST\text{-}MAX}) / V_{CELL} = (450\ \text{V}) \times (26{,}600\ \text{A}) / 5\ \text{V} = 2{,}394{,}000$ A, a value that is clearly beyond the practical realm. Consequently, multiple NaOH electrolytic cells have to be constructed and electrically connected in series to be supplied by a cell current $I_{CELL} = 96{,}500$ A that corresponds to approximately one mole of electrons supplied per second [43]. The output voltage of the PWM DC-DC converter is then calculated as $V_{OUT} = (V_{ST\text{-}MAX} \times I_{ST\text{-}MAX}) / I_{CELL} = (450\ \text{V}) \times (26{,}600\ \text{A}) / 96{,}500\ \text{A} = 124$ V. The number of NaOH electrolytic cells that must be connected in series to be supplied by a single PWM DC-DC converter is calculated as $V_{OUT} / V_{CELL} = 124\ \text{V} / 5\ \text{V} \approx 25$ cells. Therefore, the self-contained sodium (Na) metal production plant has a total of 50 NaOH electrolytic cells with 2 sets of 25 cells electrically connected in series and supplied by two identical PWM DC-DC converters that function as voltage step down converters to transform the output voltage of the solar tower V_{ST}, which represents a DC input voltage to the converter unit $V_{IN} = V_{ST\text{-}MAX} \approx 450$ V, to a constant DC output voltage $V_{OUT} = 124$ V with high efficiency, while controlling the PWM DC-DC converter input voltage V_{IN}, and thereby output voltage V_{ST} of the solar tower PV device panels to maintain their operation near the MPP.

The design of voltage step down PWM DC-DC converters that have a fixed output voltage V_{OUT} and control the input voltage V_{IN} that is variable, has been an active topic of research in the context of photovoltaic power systems applications [94, 95]. The requirements of the present application however, entail a specialized type of large scale direct photovoltaic power conversion for chemical electrolysis not hitherto contemplated or addressed in the scientific/industrial literature. The design for the voltage step down PWM DC-DC converter with a fixed output voltage V_{OUT} and variable input voltage V_{IN} meant to supply the NaOH electrolytic cells is shown in **Figure 7** to consist of a multiphase converter topology, with synchronous voltage step down converter circuits connected in parallel.

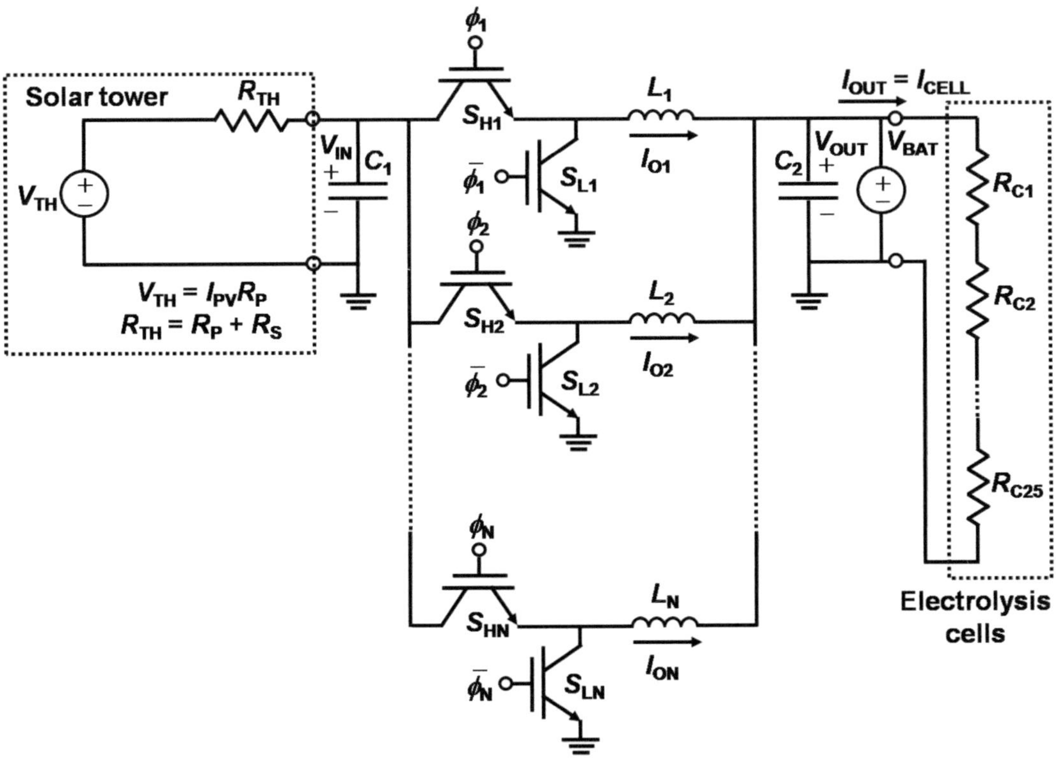

Figure 7. Multiphase voltage step down PWM DC-DC converter power supply for NaOH electrolytic cells.

The synchronous parallel multiphase voltage step down PWM DC-DC converter power supply shown in **Figure 7** offers the inherent advantage of allowing the large load current $I_{OUT} = I_{CELL} = 96{,}500$ A to be split among the phases of the converter to match the current carrying capacity of the individual electronic switches, that in practice would consist of power insulated gate bipolar transistors (IGBTs) such as the Model 1MBI3600U4D-120, manufactured by the Fuji Electric Company with a maximum rated collector-emitter voltage and collector-emitter current given as $V_{CE} = 1{,}200$ V (at $T_C = 25$ °C) and $I_{CE} = 3{,}600$ A (at $T_C = 80$ °C), respectively. It is possible to use 64 such IGBT devices in up to $N_\phi = 32$ phases of the multiphase PWM DC-DC converter with 2 IGBT devices per phase as shown in **Figure 7**, to deliver the required current $I_{OUT} = I_{CELL} = 96{,}500$ A to the 25 series connected NaOH electrolytic cells indicated as having resistances R_{C1}, R_{C2}, ..., R_{C25}, without exceeding the electrical ratings of the solid state switches. Using a multiphase converter topology with synchronous voltage step down converter circuits connected in parallel, also allows the current $I_{CELL} = 96{,}500$ A to be split among the 32 inductors L_1–L_{32}, present in each of the $N_\phi = 32$ phases, thereby not requiring one single large inductor to transmit the full current I_{OUT}, supplied by the PWM DC-DC converter to the

http://dx.doi.org/10.5772/67597

load comprising the electrolytic cells. Current sharing occurs between the synchronous voltage step down converter circuits connected in parallel according to Eq. (30).

$$I_{O1} + I_{O2} + \ldots + I_{ON} = I_{OUT} \tag{30}$$

Assuming the ideal case that current sharing between synchronous parallel voltage step down converter circuits is equal, then $I_{O1} = I_{O2} = \ldots = I_{ON} = I_{OUT} / N_{\Phi}$ where $I_{ON} = I_{OUT} / N_{\Phi} =$ (96,500 A / 32) = 3,016 A transmitted per phase. Thus, each of the 32 inductors L_1–L_{32}, can be designed for a current load I_{ON} = 3,016 A, a value well within the capabilities of inductor manufacturers. The output voltage of the multiphase voltage step down PWM DC-DC converter can be set to a fixed value using a utility scale battery, and thus $V_{BAT} = V_{OUT}$ = 124 V. The input voltage V_{IN}, of the multiphase voltage step down PWM DC-DC converter power supply for electrolytic cells shown in **Figure 7**, that also corresponds to the output voltage V_{ST}, of the solar tower PV device array has to be controlled reliably to ensure that the PV device panel array operates at or near its maximum power point (MPP).

The synchronous parallel multiphase voltage step down PWM DC-DC converter power supply in **Figure 7** with N_{Φ} = 32 phases, constitutes a complicated nonlinear dynamical system that can be challenging to model, control and analyze accurately to ensure stable operation under wide ranging conditions [100]. The closed loop control system using voltage control and a single feedback loop that can be applied to the present converter is shown in **Figure 8**.

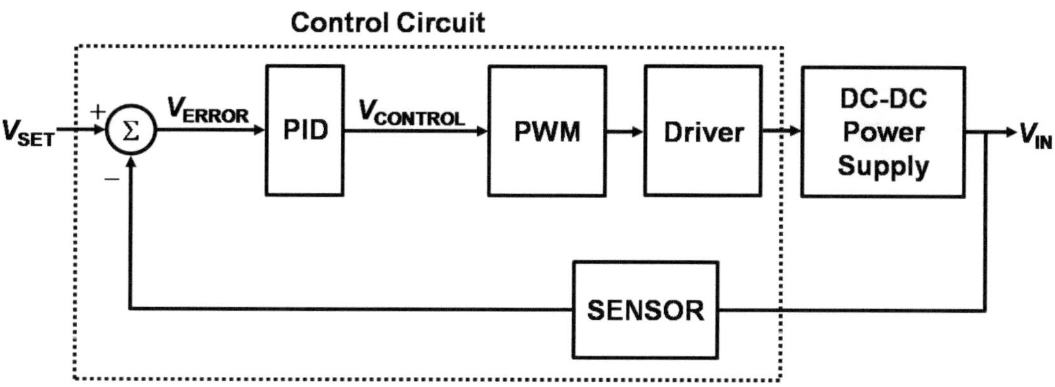

Figure 8. Closed loop control system for the multiphase voltage step down PWM DC-DC converter.

The control system for the multiphase voltage step down PWM DC-DC converter shown in **Figure 8** consists of a sensor that senses the process parameter to be controlled, namely, the input voltage V_{IN} to the PWM DC-DC converter that is also the output voltage V_{ST}, of the solar tower PV device panel array. The error amplifier compares the scaled process parameter V_{IN} to the set point value V_{SET} and computes the difference as an error signal V_{ERROR}. A proportional, integrator, differentiator (PID) circuit processes the error signal V_{ERROR}, and accordingly generates a control signal $V_{CONTROL}$, that is supplied to a pulse width modulation (PWM) signal generation circuit to produce the control signals that have the correct frequency, duty cycle and phase shift. The PWM signal to the IGBT switches has to be replicated N_{Φ} = 32 times and phase shifted by $\phi_{Shift} = 360° / N_{\phi} = 11.25°$ using a driver circuit to control all the synchronous voltage step down converters connected in parallel in **Figure 7**. In **Figure 9**, the electronic circuit schematics for some of the different control blocks in **Figure 8** are shown.

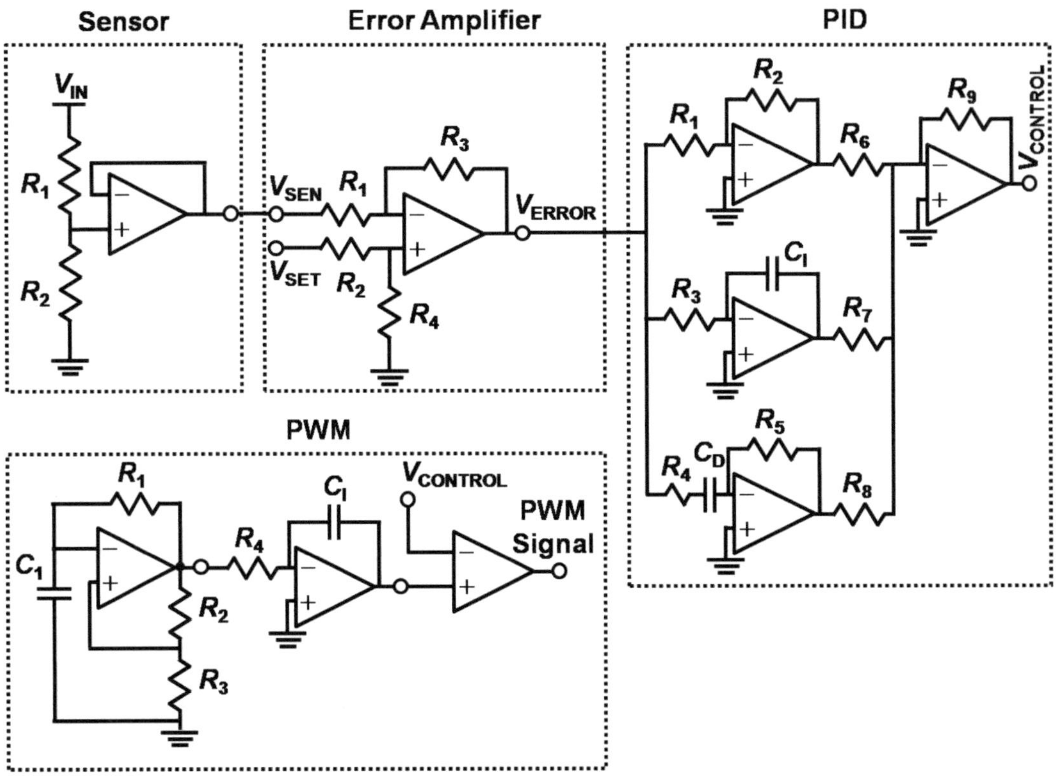

Figure 9. Electronic circuit schematics of the control system for the multiphase voltage step down PWM DC-DC converter.

The electronic circuits shown in **Figure 9** are representative of the functional blocks of the PWM DC-DC converter control system. The sensor circuit can consist of a resistive voltage divider with a unity gain op-amp buffer that senses the voltage V_{IN}, at the input of the PWM DC-DC converter according to Eq. (31).

$$V_{SEN} = V_{IN} \times (R2 / R1 + R2) \quad (31)$$

The error amplifier subtracts the input voltage V_{SEN}, scaled by the resistive voltage divider of the sensor circuit, from the set point reference voltage V_{SET}, to calculate an error voltage V_{ERROR}, according to Eq. (32).

$$V_{ERROR} = (R3 / R1) \times (V_{SET} - V_{SEN}) \text{ where } R1 = R2 \text{ and } R3 = R4 \quad (32)$$

The PID unit receives the error voltage V_{ERROR}, as an input and produces a control voltage $V_{CONTROL}$, given as Eq. (33).

$$V_{\mathrm{CONTROL}}(t) = G_{\mathrm{P}}\left(V_{\mathrm{ERROR}}(t)\right) + G_{\mathrm{I}}\int_{0}^{t} V_{\mathrm{ERROR}}(t)dt + G_{\mathrm{D}}\frac{dV_{\mathrm{ERROR}}(t)}{dt} \tag{33}$$

In Eq. (33), the terms G_P, G_I and G_D represent the gains of the proportional, integrator and differentiator circuits, respectively that can be tuned as needed to achieve optimal control characteristics for the PWM DC-DC converter. There exist myriad other ways of controlling the PWM DC-DC converter using only proportional (P) and integrator (I) control functions without the differentiator (D) circuit for example, however, the control system shown in **Figures 8** and **9**, represents a robust and reliable type of control of the PWM DC-DC converter. The PWM unit is shown to consist of a fixed frequency square wave voltage oscillator with an op-amp integrator circuit that converts the square wave into a triangular wave ramp signal which is then compared to the control voltage V_{CONTROL}, from the PID circuit using a comparator, to generate the PWM signal for the IGBT electronic switches of the PWM DC-DC converter. The driver circuit that generates the $N_{\Phi} = 32$ copies of the PWM signal from **Figure 8**, with each signal and its complement phase shifted by $\phi_{\mathrm{Shift}} = 11.25°$ for all 64 IGBT switches in the converter is not shown in **Figure 9**. It could be implemented however, using all digital logic. The control system allows the PWM DC-DC converter to be effectively controlled by setting just one parameter, namely, the reference voltage V_{SET}, to a value that corresponds with the maximum power point (MPP) output of the solar tower PV device panel array to transmit the maximum power available from the PV devices to the NaOH electrolytic cells.

It is possible to gain insight into the operation of the synchronous parallel multiphase voltage step down PWM DC-DC converter with attached solar tower PV device panel array, from the most common modeling approach using small signal analysis based on state space averaging [101]. The open loop transfer function $G_V(s)$, of the voltage step down PWM DC-DC converter that yields the small signal response of the input voltage process parameter $V_{\mathrm{IN}} = V_{\mathrm{ST}}$ to the duty cycle control variable d, can be calculated straightforwardly by making the simplifying assumption, namely, that the power supply has a single phase ($N_{\Phi} = 1$) rather than $N_{\Phi} = 32$ phases in parallel as depicted in **Figure 7**. The voltage step down PWM DC-DC converter from **Figure 7** with only a single phase ϕ_1, can be characterized by two pairs of differential equations that describe the current I_{O1}, flowing through the inductor L_1, and the voltage V_{IN}, across the input capacitor C_1, during the two distinct states of the converter when the switch S_{H1} is closed and open. The pair of differential equations corresponding to the switch S_{H1} being closed are given by Eqs. (34) and (35).

$$L_1\frac{dI_{\mathrm{O1}}}{dt} = V_{\mathrm{IN}} - V_{\mathrm{OUT}} \tag{34}$$

$$C_1\frac{dV_{\mathrm{IN}}}{dt} = \frac{V_{\mathrm{TH}} - V_{\mathrm{IN}}}{R_{\mathrm{TH}}} - I_{\mathrm{O1}} \tag{35}$$

The pair of differential equations corresponding to the switch S_{H1} being open are given by Eqs. (36) and (37).

$$L_1 \frac{dI_{O1}}{dt} = -V_{OUT} \tag{36}$$

$$C_1 \frac{dV_{IN}}{dt} = \frac{V_{TH} - V_{IN}}{R_{TH}} \tag{37}$$

The state space form of the above four differential equations describing the PWM DC-DC converter having just a single phase ϕ_1, is given as Eqs. (38) and (39) for S_{H1} being closed and open, respectively.

$$\frac{d}{dt}\begin{bmatrix} I_{O1} \\ V_{IN} \end{bmatrix} = \begin{bmatrix} 0 & \dfrac{1}{L_1} \\ -\dfrac{1}{C_1} & -\dfrac{1}{C_1 R_{TH}} \end{bmatrix} \begin{bmatrix} I_{O1} \\ V_{IN} \end{bmatrix} + \begin{bmatrix} -\dfrac{1}{L_1} & 0 \\ 0 & \dfrac{1}{C_1 R_{TH}} \end{bmatrix} \begin{bmatrix} V_{OUT} \\ V_{TH} \end{bmatrix} \tag{38}$$

$$\frac{d}{dt}\begin{bmatrix} I_{O1} \\ V_{IN} \end{bmatrix} = \begin{bmatrix} 0 & 0 \\ 0 & -\dfrac{1}{C_1 R_{TH}} \end{bmatrix} \begin{bmatrix} I_{O1} \\ V_{IN} \end{bmatrix} + \begin{bmatrix} -\dfrac{1}{L_1} & 0 \\ 0 & \dfrac{1}{C_1 R_{TH}} \end{bmatrix} \begin{bmatrix} V_{OUT} \\ V_{TH} \end{bmatrix} \tag{39}$$

The Eq. (38), has the form $\dot{x} = A_1 x + B_1 u$, and Eq. (39) has the form $\dot{x} = A_2 x + B_2 u$. The averaged state space equation can be expressed in a similar form $\dot{x} = Ax + Bu$, where the A and B matrices are calculated based on the duty cycle d, of the switch S_{H1}, the matrices A_1, B_1, A_2, B_2 and are given as $A = d \cdot A_1 + (1 - d) \cdot A_2$ and $B = d \cdot B_1 + (1 - d) \cdot B_2$. The vector x, in the averaged state space equation contains the average values of the state variables $<I_{O1}>$ and $<V_{IN}>$ and the vector u, contains the input variables V_{OUT} and V_{TH} that are considered as DC values.

When all transients in the PWM DC-DC converter have stabilized and steady state operation is achieved, then $\dot{x} = 0$, thus the averaged state space equation can be used to express the DC steady state equations given as Eqs. (40) and (41).

$$\dot{x} = 0 = AX + BU \tag{40}$$

$$X = \begin{bmatrix} I_{O1} \\ V_{IN} \end{bmatrix} = -A^{-1}BU = -\frac{L_1 C_1}{D^2} \begin{bmatrix} -\dfrac{1}{C_1 R_{TH}} & -\dfrac{D}{L_1} \\ \dfrac{D}{C_1} & 0 \end{bmatrix} \begin{bmatrix} -\dfrac{1}{L_1} & 0 \\ 0 & \dfrac{1}{C_1 R_{TH}} \end{bmatrix} \begin{bmatrix} V_{OUT} \\ V_{TH} \end{bmatrix} \tag{41}$$

Calculating out Eq. (41) yields the DC value results given in Eqs. (42) and (43) for I_{O1} and V_{IN}.

$$I_{O1} = \frac{V_{TH}}{R_{TH} D} - \frac{V_{OUT}}{R_{TH} D^2} \tag{42}$$

$$V_{IN} = \frac{V_{OUT}}{D} \tag{43}$$

The result in Eq. (43) allows to calculate the DC value duty cycle D, needed to maintain the input voltage V_{IN}, of the PWM DC-DC converter that is equal to the output voltage of the solar tower $V_{ST} = V_{IN} = 450$ V. Thus, the duty cycle $D = V_{OUT} / V_{IN} = 124$ V / 450 V = 0.28.

http://dx.doi.org/10.5772/67597

The linear model of the open loop transfer function $G_V(s)$, for the voltage step down PWM DC-DC converter at the operating point that can be used to evaluate small signal variations in the duty cycle control variable d used to control the voltage V_{IN}, can be developed by adding a small signal AC perturbation to the DC value D, of the duty cycle. Since a positive increase of the duty cycle d, causes a decrease in V_{IN} as confirmed by Eq. (43), it becomes convenient for modeling purposes to express the duty cycle in terms of a new variable $d' = 1 - d = D' + \hat{d}'$, where D' is the DC value of d' and $\hat{d}'$ is the small signal AC perturbation. The linear model of the open loop transfer function for the PWM DC-DC converter provides the response to the small signal AC perturbations $\hat{d}'$ around the DC value D'. The averaged state space equation can be formulated in terms of the new variable d' according to Eq. (44).

$$\dot{x} = \Big((1-d')A_1 + (d')A_2\Big)x + \Big((1-d')B_1 + (d')B_2\Big)u \tag{44}$$

In Eq. (44), the vector $x = X + \hat{x}$ and vector $u = U + \hat{u}$, where each vector comprises the DC value and a small signal perturbation. The Eq. (44) can be expanded as shown in Eq. (45).

$$\begin{aligned}\frac{d(X+\hat{x})}{dt} &= \Big((1-D'-\hat{d}')A_1 + (D'+\hat{d}')A_2\Big)(X+\hat{x}) + \Big((1-D'-\hat{d}')B_1 + (D'+\hat{d}')B_2\Big)(U+\hat{u}) \\ &= XA_1 + XD'(-A_1+A_2) + X\hat{d}'(-A_1+A_2) + \hat{x}A_1 + \hat{x}D'(-A_1+A_2) + \hat{x}\hat{d}'(-A_1+A_2) \\ &\quad + UB_1 + UD'(-B_1+B_2) + U\hat{d}'(-B_1+B_2) + \hat{u}B_1 + \hat{u}D'(-B_1+B_2) + \hat{u}\hat{d}'(-B_1+B_2)\end{aligned} \tag{45}$$

In Eq. (45), discarding the DC and nonlinear terms yields Eq. (46).

$$\begin{aligned}\frac{d\hat{x}}{dt} &= \hat{x}A_1 + \hat{x}D'(-A_1+A_2) + \hat{d}'\Big((-A_1+A_2)X + (-B_1+B_2)U\Big) \\ &= \hat{x}A + \hat{d}'\Big((-A_1+A_2)X + (-B_1+B_2)U\Big)\end{aligned} \tag{46}$$

Taking the Laplace transform of the averaged state space equation which has the form $\dot{x} = Ax + Bu$, yields the result in Eq. (47).

$$sX(s) - x(0) = AX(s) + BU(s) \tag{47}$$

In Eq. (47), $x(0)$ represents the initial value of the state vector in the time domain. For calculating the transfer function it is assumed that $x(0) = 0$. Solving Eq. (47) for $X(s)$ yields the result in Eq. (48).

$$X(s) = (sI_n - A)^{-1}x(0) + (sI_n - A)^{-1}BU(s) \tag{48}$$

Applying the Laplace transform to Eq. (46), yields the result given in Eq. (49).

$$s\hat{x}(s) = \hat{x}(s)A + \hat{d}'(s)\Big((-A_1+A_2)X + (-B_1+B_2)U\Big) \tag{49}$$

Solving Eq. (49) for $\hat{x}(s)$ yields the result in Eq. (50).

$$\hat{x}(s) = (s\begin{bmatrix} 1 & 0 \\ 0 & 1 \end{bmatrix} - A)^{-1}\hat{d}'(s)\Big((-A_1 + A_2)X + (-B_1 + B_2)U\Big) \tag{50}$$

The open loop transfer function given as $G_V(s) = \hat{x}(s)/\hat{d}'(s)$ of the PWM DC-DC converter can be obtained by calculating out Eq. (50) to yield the result in Eq. (51).

$$G_V(s) = \frac{R_{TH}(V_{IN}D + sL_1I_{O1})}{s^2L_1C_1R_{TH} + sL_1 + D^2R_{TH}} \tag{51}$$

The analysis to yield the open loop transfer function for the multiphase voltage step down PWM DC-DC converter with $N_{\Phi} = 32$ phases requires creating a map that contains all the state space equations describing all possible states of the 64 IGBT switches during a period of duration T_P, corresponding to a cycle of operation of the PWM DC-DC converter. For $N_{\Phi} = 32$ phases, there will be in effect 32 switching events occurring per cycle of operation with a phase shift $\phi_{Shift} = 11.25°$. Thus, there will exist 32 state space equations of the form given in Eq. (52).

$$\dot{x} = A_i x + B_i u \tag{52}$$

In Eq. (52), the matrices A_i and B_i describe the PWM DC-DC converter in the subinterval of duration t_i between switchings of the parallel phases where $\sum_{i=1}^{N_{\Phi}} t_i = T_P$. The solutions from Eq. (52) can be stacked to yield a discrete time equation valid over an entire period T_P that effectively represents a map for the multiphase PWM DC-DC converter [100]. The state space averaging method used to obtain the open loop transfer function of the PWM DC-DC converter with a single phase given in Eq. (51), can also be applied to calculate the open loop transfer function for the PWM DC-DC converter with $N_{\Phi} = 32$ phases in parallel shown in **Figure 7** [101]. An analysis of the present $N_{\Phi} = 32$ phase, closed loop multiphase PWM DC-DC converter including stability analysis, is highly complex especially when considering parametric variations among the parallel voltage step down converter circuits, and will be described in future work. It is also possible, to operate the multiphase PWM DC-DC converter in an open loop control mode, where a plant operator monitors and controls the electrical resistance of the electrolytic cells, and manually adjusts the duty cycle of the PWM DC-DC converter to deliver maximum power to the electrolytic cells. The plant operator balances the currents I_{O1}–I_{O32}, flowing in each phase of the PWM DC-DC converter by manually adjusting the amplitudes of the PWM control signals supplied to the gate terminals of the IGBT switches.

4. Sodium metal production plant operating characteristics

The solar cycle described in Section 2.2, the electrical design of the solar tower described in Section 3.1, and the electrical design of the sodium hydroxide (NaOH) electrolysis plant described in Section 3.2, entail that the voltage step down pulse width modulated (PWM) DC-DC converter supplying electricity from the solar tower PV device panel array directly to the NaOH electrolytic cells has to be operated according to a precise protocol.

Prior to sunrise occurring, the sodium hydroxide (NaOH) electrolytic cells are replenished to capacity with concentrated aqueous $NaOH_{(aq)}$ solution from the storage tanks shown in **Figure 2**, that are located adjacent to the Q-type metal building that houses electrical switch gear, voltage step down (PWM) DC-DC power converter units, the sodium hydroxide (NaOH) electrolytic cells, sodium (Na) metal packaging unit and chlorine (Cl_2) gas separation and bottling unit. As sunrise commences, the solar tower photovoltaic (PV) device panel array begins generating electric power. The voltage step down PWM DC-DC converter supplies the electric power from the solar tower PV device panel array to the electrolytic cells at a fixed voltage given as $V_{BAT} = V_{OUT} = 124$ V. Before NaOH electrolysis and Na metal production can begin, the liquid aqueous $NaOH_{(aq)}$ has to be heated to evaporate all of the water ($H_2O_{(l)}$) content and fuse or melt the remaining anhydrous solid $NaOH_{(s)}$. The presence of water or moisture in the NaOH being electrolyzed can significantly reduce Na metal yield due to side reactions occurring between the Na metal and residual H_2O. The sodium hydroxide melts into a liquid state at $T_f = 594 \pm 2$ K [102]. Therefore, electric power delivered by the solar tower initially at sunrise, has to be transmitted to electric heating elements for evaporating the water ($H_2O_{(l)}$) from the aqueous $NaOH_{(aq)}$ and to raise the temperature of the fused anhydrous $NaOH_{(l)}$ to the proper temperature for electrolysis. Once the proper temperature of the fused anhydrous $NaOH_{(l)}$ is reached, current from the solar tower can be transmitted to the cathode and anode electrodes of the 25 electrolytic cells electrically connected in series, to begin production of Na metal according to Eqs. (2) and (3).

The electric current is supplied to the 25 electrolytic cells electrically connected in series, by the synchronous parallel multiphase voltage step down PWM DC-DC converter at a fixed output voltage set by the utility scale battery given as $V_{BAT} = V_{OUT} = 124$ V. The current $I_{OUT} = I_{CELL}$, supplied by the PWM DC-DC converter from the solar tower to the electrolytic cells is not constant throughout the day, increasing gradually from sunrise until the sun reaches its zenith, and subsequently decreasing gradually as the sun tracks west and eventually sets. To use the electric power generated by the solar tower PV device panel array most efficiently for Na metal production, it is necessary to configure the electrolytic cells as a variable electric load, wherein their electrical resistance can decrease gradually as the sun arcs toward zenith allowing I_{CELL} to increase as more electric power is generated by the solar tower for electrolysis and subsequently, the electrical resistance of the electrolytic cells can begin to increase gradually as the sun passes beyond the zenith toward sunset, when I_{CELL} must decrease as less electric power is generated by the solar tower for electrolysis.

The sodium (Na) metal production plant can effectively be controlled using only two adjustable parameters, including the set point reference voltage V_{SET}, that controls the input voltage V_{IN}, of the PWM DC-DC converter that is equal to the output voltage of the solar tower $V_{IN} = V_{ST}$, and the electrical resistance of the NaOH electrolytic cells. The set point reference voltage V_{SET} can function either as a coarse or fine adjustment for the current supplied by the solar tower PV device panel array to the electrolytic cells at the fixed voltage $V_{BAT} = V_{OUT} = 124$ V. The electrical resistance of the electrolytic cells naturally increases slowly over time as the fused $NaOH_{(l)}$ reactant is decomposed by the current flowing between the electrodes of the cell according to Eqs. (2) and (3). The electrolytic cells can be designed for example, with a controllable electrical resistance that can remain relatively constant (or increase or decrease as needed) even as the fused $NaOH_{(l)}$ reactant is consumed, by varying the spacing between the anode and cathode electrodes, where the electrodes can be moved closer together to compensate the loss of reactant volume as it is consumed in the cell.

It is possible to calculate the quantity of Na metal produced throughout the year by the self-contained sodium (Na) metal production plant sited in the different geographic locations given in **Table 2**, based on the hours of daylight and the prevailing air mass conditions. It is not necessary to specify a detailed design for the NaOH electrolytic cell to generate an accurate daily estimate of Na metal production yield throughout the year, if it is assumed that maximum electric power available from the solar tower PV device panel array can always be supplied to the electrolytic cells by appropriately controlling the electrical resistance of the NaOH electrolytic cells together with the set point reference voltage V_{SET} of the PWM DC-DC converter. It is assumed that electrolysis of NaOH, and thereby Na metal production can only occur if the available current supplied by the solar tower PV device panel array has a minimum threshold value of $I_{ST} = 3{,}000$ A. In **Figure 10**, the production yield of Na metal is calculated for each day of the hypothetical year 2015, for electrolysis of pure NaOH according to Eq. (2), for the four geographic locations listed in **Table 2**, using the solar position algorithm (SPA) described in *Solar position algorithm for solar radiation applications* written by Reda & Andreas in 2004, that is a refined algorithm based on the book, *The Astronomical Algorithms* written by Meeus in 1998, and is presently regarded as the most accurate [73, 74]. The SPA allows the solar zenith θ_{sz}, and azimuth γ_s, angles to be calculated with uncertainties of $\pm 0.0003°$ in the range between -2000 to 6000 years.

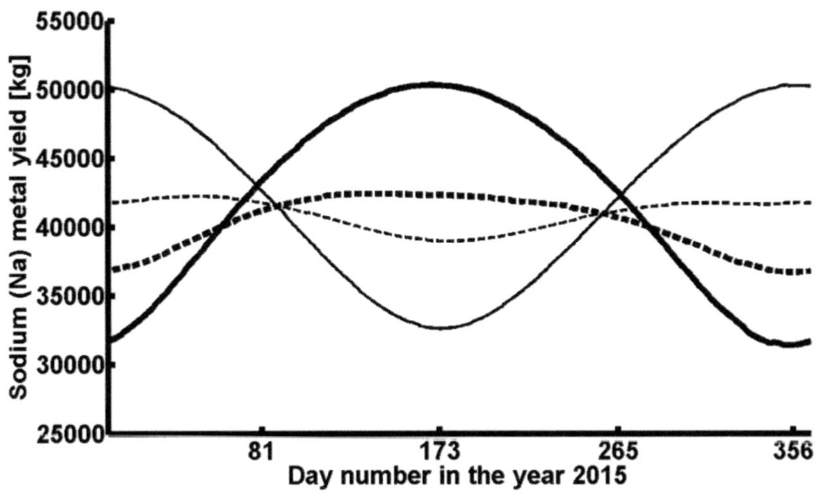

Figure 10. Calculated sodium (Na) metal daily production yields throughout the hypothetical year 2015, for El Paso, Texas (thick solid), Alice Springs, Australia (thin solid), Bangkok, Thailand (thick dash) and Mbandaka, DRC (thin dash).

The calculation in **Figure 10** provides the expected daily sodium (Na) metal production yield under the assumption that the energy conversion efficiency $\eta_{PV} = 90\%$ for the solar tower PV device panel array and furthermore, current is only transmitted to the electrolytic cells when the solar tower PV device panel array receives sufficient solar irradiance to produce the minimum threshold value of current $I_{ST} = 3{,}000$ A. The calculation in **Figure 10**, uses the SPA algorithm to determine the solar zenith angle θ_{sz}, throughout the day from sunrise to sunset for each day of the hypothetical year 2015, and calculates the air mass (AM) using Eq. (18). The direct normal total (spectrally integrated) solar irradiance incident onto the solar tower PV device panel array is calculated in turn, using Eq. (19) in a real time manner, as the solar position and air mass change throughout the day for each day of the year. The results from **Figure 10**, show that consistent quantities of Na metal are produced throughout the year when the self-contained sodium (Na) metal production plant is located as near as possible to the equator where the length of the day is the most uniform. Further away from the equator, the variability in the length of the day increases however, even at a latitude $\varphi = +31.8°$, in El Paso, Texas, the variability in the length of the day is not so significant as to render Na metal production uneconomical during the winter months when the daylight interval becomes reduced. The mean daily sodium (Na) metal production yields for the hypothetical year 2015 are calculated as 41,998 kg/day for El Paso, 41,884 kg/day for Alice Springs, 40,281 kg/day for Bangkok and 40,947 kg/day for Mbandaka, assuming a mean annual aerosol optical thickness value $k_a = 0.1$ reflecting clear days and uninterrupted Na production for all 365 days of the year. Notwithstanding the variability in daily Na metal production throughout the year, El Paso, Texas has the highest elevation above mean sea level when compared with Alice Springs, Bangkok and Mbandaka as shown in **Table 2**, resulting in an increased solar irradiance incident onto the PV device panel array and thus, the highest mean daily Na metal production. A conservative estimate for the true mean daily sodium (Na) metal production yield for the year might be $m_{Na} = 30{,}000$ kg/day of Na metal, as a consequence of nonproductive days due to inclement weather and required plant maintenance. The results in **Figure 10**, clearly demonstrate that the scalable, self-contained solar powered electrolytic sodium (Na) metal production plant can be constructed almost anywhere on earth and especially in the southwestern region of the U.S.A., to achieve a hydrogen ($H_{2(g)}$) fuel, sustainable, closed clean energy cycle.

5. Logistics of sodium hydroxide and sodium metal

For the full benefits of the hydrogen ($H_{2(g)}$) fuel based sustainable clean energy economy to be realized, it is essential to overcome the logistical problems inherent with $H_{2(g)}$ fuel. The $H_{2(g)}$ fuel possesses the unique characteristic that it can be combusted directly inside an internal combustion engine (ICE) to produce useful work without emission of carbon dioxide (CO_2) or sulfur oxides (SO_X) and with minimal emission of nitrogen oxides (NO_X). It can also be converted to electricity directly in a fuel cell to produce useful work with substantially higher efficiency than in an ICE. Regardless of how the versatile $H_{2(g)}$ fuel is applied to produce useful work, it is essential to provide safe, reliable and economical logistics for its use. The major drawback of $H_{2(g)}$ fuel remains the difficulty of direct storage. Fortunately, the element sodium (Na) positioned two rows below hydrogen in Group I of the periodic table of elements, is sufficiently electropositive to be capable of chemically releasing the $H_{2(g)}$ fuel stored in either ordinary salinated (sea) or desalinated (fresh) water (H_2O) over a wide temperature range [37]. Sodium (Na) is also sufficiently abundant in nature in the form of sodium chloride (NaCl) in seawater, to make its use economical for $H_{2(g)}$ fuel generation [103]. Therefore, sodium (Na) metal and the sodium hydroxide (NaOH) byproduct resulting from the $H_{2(g)}$ fuel producing chemical reaction in Eq. (1), constitute the ideal intermediate materials needed to render $H_{2(g)}$ into a practical and usable fuel by storing the sun's radiated energy as Na metal.

A hydrogen ($H_{2(g)}$) fuel clean energy economy based on a sustainable, closed clean energy cycle that uses sodium (Na) metal recovered by electrolysis from sodium hydroxide (NaOH) as a means of storing the sun's radiant energy collected during daytime hours, provides numerous benefits including safe, reliable and economical logistics. The scalable, self-contained sodium (Na) metal production plant that stores the sun's radiant energy in sodium (Na) metal, can be constructed in almost any geographic location on earth benefitting from ample solar irradiance and clear weather all year. In the U.S.A., the arid, southwestern desert region offers the requisite conditions, including sufficient undeveloped land area to construct scalable, self-contained solar powered electrolytic sodium (Na) metal production plants by the thousands. Using the southwestern desert region that includes West Texas, New Mexico, Arizona and Southern California as a hub for solar powered sodium (Na) metal production by electrolysis of sodium hydroxide (NaOH), it is possible to develop sufficient Na metal production capacity based on the scalable, self-contained sodium (Na) metal production plant described, to meet the U.S.A.'s energy needs for motor vehicle transport and for broader clean electric power applications.

The physical and chemical properties of sodium (Na) metal and sodium hydroxide (NaOH) render these materials ideal from an operational logistical standpoint. The sodium (Na) metal is a solid at room temperature and therefore has negligible vapor pressure. As a result, Na metal can be stored almost indefinitely in hermetically sealed packaging that can be opened much as a sardine can, only when the Na metal must be loaded into a hydrogen generation apparatus to react with either salinated (sea) or desalinated (fresh) water (H_2O) according to Eq. (1), to produce hydrogen ($H_{2(g)}$) fuel on demand [37]. The sodium hydroxide (NaOH) byproduct of the hydrogen producing chemical reaction in Eq. (1), is also a solid at room temperature in its pure form and has negligible vapor pressure. The NaOH is miscible with water in all proportions, enabling the aqueous $NaOH_{(aq)}$ solution to be readily transferred by pumping into and out of sealed tanks for transport by truck, rail car or pipeline to the remotely located self-contained sodium (Na) metal production plant units for reprocessing by electrolysis according to Eqs. (2) and (3), to recover the Na metal for reuse in generating $H_{2(g)}$ fuel. The $NaOH_{(aq)}$ transportation/storage tanks of the type shown in **Figure 2**, can be fiberglass or metal with a corrosion resistant internal rubber liner, and must seal hermetically to exclude ambient air that contains carbon dioxide (CO_2) which slowly degrades the $NaOH_{(aq)}$, albeit not irreversibly.

To obtain a sense for the magnitude of the logistical effort needed to produce and distribute sufficient sodium (Na) metal to fuel all of the motor vehicles in the U.S.A. while recovering the sodium hydroxide (NaOH) byproduct for reprocessing by electrolysis, it is necessary to consider the total number of vehicles in circulation. According to the Bureau of Transportation Statistics at the United States Department of Transportation (DOT), the total number of registered vehicles in the year 2013 in the U.S.A. numbered 255,876,822 [104]. The figure includes passenger cars, motorcycles, light duty vehicles, other 2-axle/4-tire vehicles, trucks with 2-axles/6-tires or more and buses. If it is further assumed that each motor vehicle on average consumes the energy equivalent of 16.2 gallons of 100 octane gasoline (2,2,4-Trimethylpentane) per week, then the corresponding quantity of $H_{2(g)}$ fuel having an equivalent heating value is given as 15.8 kg. The generation of 15.8 kg of $H_{2(g)}$ fuel according to

http://dx.doi.org/10.5772/67597

Eq. (1), requires that 361.6 kg of Na metal react with approximately 300 kg of water (H_2O) [37]. Therefore, the total quantity of Na metal consumed per week in the U.S.A. can be calculated as 255,876,822 vehicles × 361.6 kg/week = 92,525,058,835 kg/week. If each solar tower produces a mass m_{Na} = 30,000 kg/day of Na metal, then in one week the Na metal yield per solar tower will be given as 30,000 kg × 7 days = 210,000 kg/week. The number of solar towers required to meet demand for Na metal will be given as N_{ST} = 92,525,058,835 kg/ week / 210,000 kg/week = 440,596 solar towers or approximately $N_{ST} \approx$ 450,000 solar towers. While the number of solar towers required might seem very large and the task of constructing them onerous, it is in fact possible to construct the sufficient number of towers to provide Na metal for all the motor vehicles in the U.S.A. The self-contained sodium (Na) metal production plants can be constructed at a density of approximately ρ_{plant} = 30 plant units per square mile to prevent mutual shading when the towers are elevated and rotated to track the sun. The solar tower density and layout necessitate a land area given as $A_{land} = N_{ST}$ / ρ_{plant} = 450,000 / 30 = 15,000 square miles, to meet the Na metal demand for all the motor vehicles in the U.S.A. using PV device panels with an efficiency η_{PV} = 90%, and a land area A_{land} = 75,000 square miles using PV device panels with an efficiency η_{PV} = 18% that currently exist commercially. The land area required can be placed into perspective when considering that the area of the state of New Mexico is approximately 121,000 square miles and thus, there exists more than sufficient desert land area for constructing the scalable, self-contained sodium (Na) metal production plants in the U.S.A.

Our company believes that hydrogen ($H_{2(g)}$) fuel will earn an important role in motor vehicle transport applications for powering smaller 1–5 kW class secondary power fuel cells for onboard continuous recharging of battery electric vehicles (BEVs), a concept implemented successfully in the 1960s using $H_{2(g)}$ fuel stored in high pressure cylinders [23]. The concept of a smaller hydrogen fuel cell operating at a fixed power output level to continuously recharge an electric storage battery can be extended not only to motor vehicle propulsion systems but also for a broad range of clean electric power applications, including general ground transport that includes commercial trucks, trains, maritime transport as well as powering single family homes, commercial establishments and industrial enterprises. Such an approach will ultimately enable mankind to dispense with use of carbon based fossil fuels for motor vehicle transport applications and most other types of ground based electric power generation.

6. Conclusion

The technical and economic viability of a novel, scalable, self-contained solar powered electrolytic sodium (Na) metal production plant has been demonstrated for meeting the hydrogen ($H_{2(g)}$) fuel clean energy needs of the U.S.A. The scalable, self-contained sodium (Na) metal production plant uses a solar tower PV device panel array to collect and convert the sun's vast radiant energy emission produced by hydrogen fusion, into electric power that is used to recover sodium (Na) metal from sodium hydroxide (NaOH) or from a mixture of NaOH and NaCl by electrolysis. The Na metal can subsequently be reused to generate $H_{2(g)}$ fuel and NaOH byproduct by reacting with either ordinary salinated (sea) or desalinated (fresh) water (H_2O). The scalable, self-contained sodium (Na) metal production plant operation is enabled by a specially designed voltage step down PWM DC-DC converter consisting of a multiphase converter topology with up to 32 synchronous voltage step down converter circuits connected in parallel. The PWM DC-DC converter has a fixed output voltage $V_{OUT} \approx 124$ V and variable input voltage $V_{IN} = V_{ST}$, that corresponds to the output voltage of the solar tower PV device panel array and can be controlled to maintain the PV device panel array operating near the maximum power point (MPP). Each scalable, self-contained sodium (Na) metal production plant consists of two voltage step down PWM DC-DC converters, wherein each unit supplies 25 NaOH electrolytic cells, electrically connected in series, with a current $I_{CELL} = 96{,}500$ A, corresponding to approximately one mole of electrons per second. The scalable electrical design of the solar tower allows the PV device panel array to be upgraded with newer and more efficient PV device panels as they become available as a result of progress in scientific research and development. Once PV device panels with an efficiency $\eta_{PV} = 90\%$ will become available, the power output of the solar tower PV device panel array can reach $P_{ST} = 23.9$ MW that is sufficient for producing a mass quantity of approximately $m_{Na} = 30{,}000$ kg of Na metal per day from the electrolysis of NaOH. It therefore becomes possible to meet the hydrogen ($H_{2(g)}$) fuel clean energy needs of all the motor vehicles in the U.S.A. by constructing approximately 450,000 scalable, self-contained sodium (Na) metal production plant units of the type described, in the southwestern desert region of the U.S.A. that includes West Texas, New Mexico, Arizona and Southern California. If the land area needed for the scalable, self-contained sodium (Na) metal production plant units becomes scarce, then purpose built ships equipped with the Na metal production plants can be dispatched into the vast ocean expanses near the equator where high solar irradiance occurs, to convert aqueous $NaOH_{(aq)}$ stored onboard into sodium (Na) metal before returning to port.

Nomenclature

a	Length of the semi-major axis of earth's elliptical orbit around the sun	(m)
A, B	Matrices	
A_P	Photovoltaic (PV) panel area	(m^2)
A_{PA}	Photovoltaic (PV) panel array area	(m^2)
A_{land}	Land area	(mi^2)
AM	Air mass at mean sea level	
AM_a	Air mass at actual atmospheric pressure	
B_l	Solar tower branch length	(m)
$Bsec_l$	Solar tower branch section length	(m)
B_h	Solar tower branch height	(m)
d	DC-DC converter duty cycle	

http://dx.doi.org/10.5772/67597

d'	DC-DC converter duty cycle, $d' = 1 - d$	
$\hat{d}'$	DC-DC converter duty cycle small signal AC perturbation	
C	Capacitor value	(F)
D	DC-DC converter duty cycle DC value	
D'	DC-DC converter duty cycle DC value, $D' = 1 - D$	
D_S	Electrical conductor diameter	(cm)
i	Indices	
E_r°, E_o°	Standard reduction, oxidation half reaction potential	(V)
E_{ov}°	Standard overall reaction potential	(V)
E_0	Eccentricity correction for the solar constant	
G_P	Proportional circuit gain	
G_I	Integrator circuit gain	
G_D	Differentiator circuit gain	
G_V	DC-DC converter open loop voltage transfer function	
h	Hour angle	(°)
h_{PV}	Photovoltaic (PV) array elevation above mean sea level	(m)
H_{sr}	Hours between sunrise and *local solar noon*	(hours)
H_{ss}	Hours between *local solar noon* and sunset	(hours)
Irr_n	Direct normal solar irradiance	(W/m^2)
$Irr_{AM1.5D}$	ASTM direct normal AM 1.5D standard terrestrial total solar irradiance	(W/m^2)
$Irr_{AM1.5G}$	ASTM global AM 1.5G standard terrestrial total solar irradiance	(W/m^2)
I_B	Solar tower branch current	(A)
I_{Bsec}	Solar tower branch section current	(A)
I_{ST}	Solar tower output current	(A)
$I_{ST\text{-}MAX}$	Solar tower maximum output current	(A)
I_O	DC-DC converter per phase output current	(A)
I_{OUT}	DC-DC converter output current	(A)
I_{CELL}	Electrolytic cell current	(A)
I_{PV}	Photovoltaic (PV) device current	(A)
I_{SC}	Photovoltaic (PV) panel short circuit current	(A)
$I_{MPP\text{-}P}$	Photovoltaic (PV) panel maximum power point current	(A)
$I_{MPP\text{-}PA}$	Photovoltaic (PV) panel array maximum power point current	(A)
I_{CE}	IGBT collector-emitter current	(A)
k_a	Aerosol optical depth or thickness	
l_o	Vertical ozone layer depth or thickness	(cm (NTP))
L	Inductor value	(H)
m_{Na}	Sodium mass	(kg)

N_P	Number of photovoltaic (PV) panels	
$N_{P\text{-}B}$	Number of photovoltaic (PV) panels per branch	
$N_{P\text{-}Bsec}$	Number of photovoltaic (PV) panels per branch section	
$N_{B\text{-}L/R}$	Number of branches on the left or right of the solar tower	
$N_{B\text{-}ST}$	Number of branches on the solar tower	
N_ϕ	Number of phases	
N_{ST}	Number of solar towers	
N_{day}	Day number in a year from 1 to 365	
P	Absolute pressure	(Pa)
$P_{ST\text{-}L/R}$	Solar tower left or right half output power	(W) or (MW)
P_{ST}	Solar tower output power	(W) or (MW)
$P_{ST\text{-}50}$	Solar tower output power (50 towers)	(W) or (MW)
P_w	Photovoltaic (PV) panel width	(m)
P_l	Photovoltaic (PV) panel length	(m)
r_{sun}	Distance from the center of the sun to the center of the earth	(m)
R	Resistor value	(Ω)
R_P, R_S	Photovoltaic (PV) device parallel resistance, series resistance	(Ω)
R_{TH}	Thevenin equivalent resistance	(Ω)
R_C	Electrolytic cell resistance	(Ω)
S_h	Solar tower structure height	(m)
S_w	Solar tower structure width	(m)
t	Time duration	
T	Absolute temperature, ITS-90 or Celsius temperature	(K) or (°C)
T_C	Surface temperature of IC package	(°C)
T_f	Fusion temperature	(K)
T_P	Time period for a cycle	
u, U	Vector, vector DC component	
$V_{NaOH(aq)}$	Aqueous sodium hydroxide volume	(Gal)
V_{ST}	Solar tower output voltage	(V)
$V_{ST\text{-}MAX}$	Solar tower maximum output voltage	(V)
$V_{ST\text{-}DROP}$	Solar tower central column conductor voltage drop	(V)
V_{IN}	DC-DC converter input voltage	(V)
V_{OUT}	DC-DC converter output voltage	(V)
V_{BAT}	Utility scale battery voltage	(V)
V_{CELL}	Electrolytic cell voltage	(V)
V_{OC}	Photovoltaic (PV) panel open circuit voltage	(V)
$V_{MPP\text{-}C}$	Photovoltaic (PV) single cell maximum power point voltage	(V)

http://dx.doi.org/10.5772/67597

$V_{MPP\text{-}P}$	Photovoltaic (PV) panel maximum power point voltage	(V)
$V_{MPP\text{-}PA}$	Photovoltaic (PV) panel array maximum power point voltage	(V)
V_{TH}	Thevenin equivalent voltage	(V)
V_{CE}	IGBT collector-emitter voltage	(V)
V_{SEN}	DC-DC converter scaled input voltage	(V)
V_{SET}	DC-DC converter input voltage, set point reference	(V)
V_{ERROR}	DC-DC converter control circuit, error amplifier output	(V)
$V_{CONTROL}$	DC-DC converter control circuit, PID circuit output	(V)
w'	Precipitable water thickness at actual atmospheric pressure and temperature	(cm)
x, X	Vector, vector DC component	
α_s	Solar altitude or elevation angle above the observer's horizon	(°)
$\alpha_{s\text{-}max}$	Maximum solar altitude or elevation angle above the observer's horizon	(°)
ϕ_{Shift}	Phase shift	(°)
γ_s	Solar azimuth angle	(°)
Γ	Day angle	(radians)
η_{CELL}	Electrolytic cell efficiency	(%)
$\eta_{DC\text{-}DC}$	DC-DC converter efficiency	(%)
η_{PV}	Photovoltaic (PV) device panel efficiency	(%)
φ	Geographic latitude	(°)
λ	Geographic longitude	(°)
λ_e	Ecliptic longitude	(°)
λ_L	Local longitude	(°)
λ_S	Standard longitude	(°)
θ	Angle between position of earth in orbit around sun and perihelion position	(°)
θ_{sz}	Solar zenith angle	(°)
ρ_{plant}	Photovoltaic (PV) plant density	(mi^{-2})
τ_r	Transmittance by Rayleigh scattering	
τ_o	Transmittance by ozone	
τ_g	Transmittance by uniformly mixed gases	
τ_w	Transmittance by precipitable water vapor	
τ_a	Transmittance by aerosol	
R_g	Molar gas constant	8.3144621 (J/K·mol)
M_{air}	Molar mass, air	0.028964 (kg/mol)
R_{air}	Specific gas constant, air	287.06194 (J/K·kg)
e	Eccentricity of earth's elliptical orbit around the sun	0.01673
F	Faraday constant	96485.3365 (C/mol)
g_0	Gravitational acceleration near earth's surface	9.80665 (m/s^2)

Irr_0	Solar constant	1367 (W/m^2)
P_0	Standard atmospheric pressure	101325 (Pa)
P_{sun}	Power output of the sun	3.8×10^{26} (W)
P_{earth}	Power output of the sun reaching the earth	1.7×10^{17} (W)
R_{sun}	Radius of the sun	6.96×10^{8} (m)
$r_{sun\text{-}m}$	Mean distance from center of sun to center of earth	1.496×10^{11} (m)
T_0	Celsius zero point, ITS-90	273.15 (K)
T_{Eu}	Eutectic temperature of NaCl-H_2O solution	−21.2 (°C)
T_{sun}	Surface temperature of the solar black body	5800 (K)
T_{es}	Period of earth's rotation around the sun	365.24 (days)
T_{ea}	Period of earth's rotation on its axis (*mean solar day*)	86,400 (sec)
ε	Obliquity or tilt angle of earth's rotation axis	23.44 (°)
δ	Solar angle of declination	$-23.44 \le \delta \le +23.44$ (°)
η_{PVmax}	Thermodynamic efficiency limit of PV device panels	93 (%)
$\varphi_{T\text{-}CAN}$	Latitude at Tropic of Cancer	+23.44 (°)
$\varphi_{T\text{-}CAP}$	Latitude at Tropic of Capricorn	−23.44 (°)
λ_{PM}	Longitude at Greenwich Prime Meridian	0 (°)
π	Number, pi	3.14
ω_{ea}	Angular velocity of earth's rotation on its axis	7.292115×10^{-5} (rad/sec)

Author details

Alvin G. Stern

Address all correspondence to: inquiries@agstern.com

AG STERN, LLC, Newton, MA, USA

References

[1] Lutz, W., Samir, K.C., "Dimensions of global population projections: what do we know about future population trends and structures?," Philosophical Transactions of the Royal Society of London B: Biological Sciences, 365(1554), 2779–2791, (2010).

[2] Cohen, J.E., "Human Population: The Next Half Century," Science, 302(5648), 1172–1175, (2003).

[3] Cohen, J.E., "Population Growth and Earth's Human Carrying Capacity," Science, 269 (5222), 341–346, (1995).

[4] Narayan, P.K., Smyth, R., "Energy consumption and real GDP in G7 countries: New evidence from panel cointegration with structural breaks," Energy Economics, 30(5), 2331–2341, (2008).

[5] Soytas, U., Sari, R., "Energy consumption and income in G-7 countries," Journal of Policy Modeling, 28(7), 739–750, (2006).

[6] Stern, D.I., "Energy and economic growth in the USA: A multivariate approach," Energy Economics, 15(2), 137–150, (1993).

[7] Dignon, J., "NO_X and SO_X emissions from fossil fuels: A global distribution," Atmospheric Environment. Part A. General Topics, 26(6), 1157–1163, (1992).

[8] Gustafsson, Ö., Krusa, M., Zencak, Z., Sheesley, R.J., Granat, L., Engström, E., Praveen, P.S., Rao, P.S.P., Leck, C., Rodhe, H., "Brown Clouds over South Asia: Biomass or Fossil Fuel Combustion?," Science, 323(5913), 495–498, (2009).

[9] Mukhopadhyay, K., Forssell, O., "An empirical investigation of air pollution from fossil fuel combustion and its impact on health in India during 1973–1974 to 1996–1997," Ecological Economics, 55(2), 235–250, (2005).

[10] Streets, D.G., Waldhoff, S.T., "Present and future emissions of air pollutants in China: SO_2, NO_X, and CO," Atmospheric Environment, 34(3), 363–374, (2000).

[11] Reddy, M.S., Venkataraman, C., "Inventory of aerosol and sulphur dioxide emissions from India: I-Fossil fuel combustion," Atmospheric Environment, 36(4), 677–697, (2002).

[12] Chakraborty, N., Mukherjee, I., Santra, A.K., Chowdhury, S., Chakraborty, S., Bhattacharya, S., Mitra, A.P., Sharma, C., "Measurement of CO_2, CO, SO_2, and NO emissions from coal-based thermal power plants in India," Atmospheric Environment, 42(6), 1073–1082, (2008).

[13] Chan, C.C., "The State of the Art of Electric, Hybrid, and Fuel Cell Vehicles," Proceedings of the IEEE, 95(4), 704–718, (2007).

[14] Romm, J., "The car and fuel of the future," Energy Policy, 34(17), 2609–2614, (2006).

[15] Rotering, N., Ilic, M., "Optimal Charge Control of Plug-In Hybrid Electric Vehicles in Deregulated Electricity Markets," IEEE Transactions on Power Systems, 26(3), 1021–1029, (2011).

[16] Hoffmann, B.S., Szklo, A., "Integrated gasification combined cycle and carbon capture: A risky option to mitigate CO_2 emissions of coal-fired power plants," Applied Energy, 88(11), 3917–3929, (2011).

[17] Sims, R.E.H., Rogner, H.H., Gregory, K., "Carbon emission and mitigation cost comparisons between fossil fuel, nuclear and renewable energy resources for electricity generation," Energy Policy, 31(13), 1315–1326, (2003).

[18] Rosen, M.A., "Energy- and exergy-based comparison of coal-fired and nuclear steam power plants," Exergy, An International Journal, 1(3), 180–192, (2001).

[19] Benke, A.C., "A Perspective on America's Vanishing Streams," Journal of the North American Benthological Society, 9(1), 77–88, (1990).

[20] Kosnik, L., "The potential for small scale hydropower development in the US," Energy Policy, 38(10), 5512–5519, (2010).

[21] Eberle, U., Müller, B., von Helmolt, R., "Fuel cell electric vehicles and hydrogen infrastructure: status 2012," Energy & Environmental Science, 5(10), 8780–8798, (2012).

[22] von Helmolt, R., Eberle, U., "Fuel cell vehicles: Status 2007," Journal of Power Sources, 165(2), 833–843, (2007).

[23] Kordesch, K.V., "Hydrogen-Air/Lead Battery Hybrid System for Vehicle Propulsion," Journal of The Electrochemical Society, 118(5), 812–817, (1971).

[24] Bagotsky, V.S., Fuel Cells: Problems and Solutions. Hoboken, NJ: John Wiley & Sons, pp. 7–70, 109–123, 189–239, (2009).

[25] Eberle, U., von Helmolt, R., "Sustainable transportation based on electric vehicle concepts: a brief overview," Energy & Environmental Science, 3(6), 689–699, (2010).

[26] Paster, M.D., Ahluwalia, R.K., Berry, G., Elgowainy, A., Lasher, S., McKenney, K., Gardiner, M., "Hydrogen storage technology options for fuel cell vehicles: Well-to-wheel costs, energy efficiencies, and greenhouse gas emissions," International Journal of Hydrogen Energy, 36(22), 14534–14551, (2011).

[27] Schüth, F., Bogdanovic, B., Felderhoff, M., "Light metal hydrides and complex hydrides for hydrogen storage," Chemical Communications, 20, 2249–2258, (2004).

[28] Zaluski, L., Zaluska, A., Ström-Olsen, J.O., "Hydrogenation properties of complex alkali metal hydrides fabricated by mechano-chemical synthesis," Journal of Alloys and Compounds, 290(1–2), 71–78, (1999).

[29] Bogdanovic, B., Schwickardi, M., "Ti-doped alkali metal aluminum hydrides as potential novel reversible hydrogen storage materials," Journal of Alloys and Compounds, 253–254, 1–9, (1997).

[30] Zhang, Y., Liao, S., Xu, Y., "Highly active alkali metal hydrides; their catalytic syntheses and properties," Journal of Molecular Catalysis, 84(3), 211–221, (1993).

[31] Knowlton, R.E., "An investigation of the safety aspects in the use of hydrogen as a ground transportation fuel," International Journal of Hydrogen Energy, 9(1–2), 129–136, (1984).

[32] Hord, J., "Is hydrogen a safe fuel?," International Journal of Hydrogen Energy, 3(2), 157–176, (1978).

[33] Brown, L.F., "A comparative study of fuels for on-board hydrogen production for fuel-cell-powered automobiles," International Journal of Hydrogen Energy, 26(4), 381–397, (2001).

[34] Xu, J., Froment, G.F., "Methane Steam Reforming, Methanation and Water-Gas Shift: I. Intrinsic Kinetics," AIChE Journal, 35(1), 88–96, (1989).

[35] Baschuk, J.J., Li, X., "Carbon monoxide poisoning of proton exchange membrane fuel cells," International Journal of Energy Research, 25(8), 695–713, (2001).

[36] Si, Y., Jiang, R., Lin, J.C., Kunz, H.R., Fenton, J.M., "CO Tolerance of Carbon-Supported Platinum-Ruthenium Catalyst at Elevated Temperature and Atmospheric Pressure in a PEM Fuel Cell," Journal of The Electrochemical Society, 151(11), A1820–A1824, (2004).

[37] Stern, A.G., "Design of an efficient, high purity hydrogen generation apparatus and method for a sustainable, closed clean energy cycle," International Journal of Hydrogen Energy, 40(32), 9885–9906, (2015).

[38] Bodnar, R.J., "Revised equation and table for determining the freezing point depression of H_2O-NaCl solutions," Geochimica et Cosmochimica Acta, 57(3), 683–684, (1993).

[39] Hall, D.L., Sterner, S.M., Bodnar, R.J., "Freezing point Depression of NaCl-KCl-H_2O Solutions," Economic Geology, 83(1), 197–202, (1988).

[40] Castner, H.Y., "Process of manufacturing sodium and potassium," U.S. Patent 452,030, (1891).

[41] Downs, J.C., "Electrolytic process and cell," U.S. Patent 1,501,756, (1924).

[42] Davy, H., "The Bakerian Lecture: On Some New Phenomena of Chemical Changes Produced by Electricity, Particularly the Decomposition of the Fixed Alkalies, and the Exhibition of the New Substances Which Constitute Their Bases; And on the General Nature of Alkaline Bodies" Philosophical Transactions of the Royal Society of London, 98, 1–44, (1808).

[43] Weast, R.C., Handbook of Chemistry and Physics. Cleveland, OH: CRC Press, pp. F13, F104, F242, D141–D146, (1976–1977).

[44] Daly, C., Halbleib, M., Smith, J.I., Gibson, W.P., Doggett, M.K., Taylor, G.H., Curtis, J., Pasteris, P.P., "Physiographically sensitive mapping of climatological temperature and precipitation across the conterminous United States," International Journal of Climatology, 28(15), 2031–2064, (2008).

[45] Wilson, G.H., "The Hottest Region in the United States," Monthly Weather Review, 43 (6), 278–280, (1915).

[46] Court, A., "How Hot is Death Valley?," Geographical Review, 39(2), 214–220, (1949).

[47] Fthenakis, V., Mason, J.E., Zweibel, K., "The technical, geographical, and economic feasibility for solar energy to supply the energy needs of the US," Energy Policy, 37(2), 387–399, (2009).

[48] Renewable Resource Data Center, The National Solar Radiation Data Base (NSRDB), National Renewable Energy Laboratory, Golden, CO. U.S.A.

[49] Frosch, R.A., Gallopoulos, N.E., "Strategies for Manufacturing," Scientific American, 261(3), 144–152, (1989).

[50] Henry, C.H., "Limiting efficiencies of ideal single and multiple energy gap terrestrial solar cells," Journal of Applied Physics, 51(8), 4494–4500, (1980).

[51] Press, W.H., "Theoretical maximum for energy from direct and diffuse sunlight," Nature, 264, 734–735, (1976).

[52] Shockley, W., Queisser, H.J., "Detailed Balance Limit of Efficiency of p-n Junction Solar Cells," Journal of Applied Physics, 32(3), 510–519, (1961).

[53] Bagnall, D.M., Boreland, M., "Photovoltaic technologies," Energy Policy, 36(12), 4390–4396, (2008).

[54] Stern, A.G., "Design of high quantum efficiency and high resolution, Si/SiGe avalanche photodiode focal plane arrays using novel, back-illuminated, silicon-on-sapphire substrates," in Park, J.W., Editor. Photodiodes – World Activities in 2011. Vienna: InTech Publisher, pp. 267–312, ISBN 978-953-307-530-3, (2011).

[55] Stern, A.G., "Design of a high sensitivity emitter-detector avalanche photodiode imager using very high transmittance, back-illuminated, silicon-on-sapphire," SPIE Journal of Optical Engineering, 51(6), 063206, (2012).

[56] Stern, A.G., "Very high transmittance, back-illuminated, silicon-on-sapphire semiconductor wafer substrate for high quantum efficiency and high resolution, solid-state, imaging focal plane arrays," U.S. Patent 8,354,282, (2013).

[57] Stern, A.G., "Thin, very high transmittance, back-illuminated, silicon-on-sapphire semiconductor substrates bonded to fused silica," U.S. Patent 8,471,350, (2013).

[58] Sadok, M., Benyoucef, B., Benmedjahed, M., "Assessment of PV Modules Degradation based on Performances and Visual Inspection in Algerian Sahara," International Journal of Renewable Energy Research, 6(1), 106–116, (2016).

[59] Garg, H.P., Treatise on Solar Energy: Fundamentals of Solar Energy, Volume 1. New York: John Wiley & Sons, pp. 50–150, (1982).

[60] Sproul, A.B., "Derivation of the solar geometric relationships using vector analysis," Renewable Energy, 32(7), 1187–1205, (2007).

[61] Kalogirou, S.A., Solar Energy Engineering, Processes and Systems, 2nd edition. Waltham, MA: Academic Press, pp. 51–122, (2014).

[62] Guinot, B., Seidelmann, P.K., "Time scales: their history, definition and interpretation," Astronomy and Astrophysics, 194(1–2), 304–308, (1988).

[63] Essen, L., "Time Scales," Metrologia, 4(4), 161–165, (1968).

[64] Aoki, S., Guinot, B., Kaplan, G.H., Kinoshita, H., McCarthy, D.D., Seidelmann, P.K., "The New Definition of Universal Time," Astronomy and Astrophysics, 105(2), 359–361, (1982).

http://dx.doi.org/10.5772/67597

[65] Capitaine, N., Gontier, A.M., "Accurate procedure for deriving UT1 at a submilliarcsecond accuracy from Greenwich Sidereal Time or from the stellar angle," Astronomy and Astrophysics, 275, 645–650, (1993).

[66] Capitaine, N., Wallace, P.T., McCarthy, D.D., "Expressions to implement the IAU 2000 definition of UT1," Astronomy and Astrophysics, 406(3), 1135–1149, (2003).

[67] Moritz, H., "Geodetic reference system 1980," Journal of Geodesy, 54(3), 395–405, (1980).

[68] Malys, S., Seago, J.H., Pavlis, N.K., Seidelmann, P.K., Kaplan, G.H., "Why the Greenwich meridian moved," Journal of Geodesy, 89(12), 1263–1272, (2015).

[69] Spencer, J.W., "Fourier series representation of the position of the sun," Search, 2(5), 172, (1971).

[70] Hughes, D.W., Yallop, B.D., Hohenkerk, C.Y., "The Equation of Time," Monthly Notices of the Royal Astronomical Society, 238(4), 1529–1535, (1989).

[71] Grena, R., "Five new algorithms for the computation of sun position from 2010 to 2110," 86(5), 1323–1337, (2012).

[72] Grena, R., "An algorithm for the computation of the solar position," Solar Energy, 82(5), 462–470, (2008).

[73] Reda, I., Andreas, A., "Solar position algorithm for solar radiation applications," Solar Energy, 76(5), 577–589, (2004).

[74] Reda, I., Andreas, A., "Solar Position Algorithm for Solar Radiation Applications," NREL/TP-560-34302. Golden, CO: National Renewable Energy Laboratory, (2008).

[75] Blanco-Muriel, M., Alarcón-Padilla, D.C., López-Moratalla, T., Lara-Coira, M., "Computing the solar vector," Solar Energy, 70(5), 431–441, (2001).

[76] Michalsky, J.J., "*The Astronomical Almanac's* algorithm for approximate solar position (1950–2050)," Solar Energy, 40(3), 227–235, (1988).

[77] Young, A.T., "Air mass and refraction," Applied Optics, 33(6), 1108–1110, (1994).

[78] Iqbal, M., An Introduction to Solar Radiation. New York, NY: Academic Press, pp. 85–212, (1983).

[79] Bird, R., Hulstrom, R.L., "Direct Insolation Models," SERI/TR-335-344. Golden, CO: Solar Energy Research Institute, (1980).

[80] Bird, R., Hulstrom, R.L., "A Simplified Clear Sky Model for Direct and Diffuse Insolation on Horizontal Surfaces," SERI/TR-642-761. Golden, CO: Solar Energy Research Institute, (1981).

[81] Taylor, B.N., Thompson, A., "The International System of Units (SI)," NIST Special Publication 330. Gaithersburg, MD: National Institute of Standards and Technology, (2008).

[82] Mohr, P.J., Taylor, B.N., Newell, D.B., "CODATA recommended values of the fundamental physical constants: 2010," Reviews of Modern Physics, 84(4), 1527–1605, (2012).

[83] Mitchell, T.D., Carter, T.R., Jones, P.D., Hulme, M., New, M., "A comprehensive set of high-resolution grids of monthly climate for Europe and the globe: the observed record (1901–2000) and 16 scenarios (2001–2100). (CRU TS 2.0)," Tyndall Centre Working Paper 55, Climatic Research Unit, School of Environmental Sciences, University of East Anglia, pp. 1–25, (2004).

[84] Randel, D.L., Vonder Haar, T.H., Ringerud, M.A., Stephens, G.L., Greenwald, T.J., Combs, C.L., "A New Global Water Vapor Dataset," Bulletin of the American Meteorological Society, 77(6), 1233–1246, (1996).

[85] Tuller, S.E., "World Distribution of Mean Monthly and Annual Precipitable Water," Monthly Weather Review, 96(11), 785–797, (1968).

[86] Phokate, S., Pisutthipong, N., "The Transmittance to Direct Irradiance due to Absorption by Precipitable Water Vapor in the Atmosphere," Advanced Materials Research, 979, 7–10, (2014).

[87] Phokate, S., "The Distribution of Precipitable Water Vapor in the Atmosphere of Thailand," In Proceeding of 4th International Science, Social Science, Engineering and Energy Conference (I-SEEC), Petchburi, Thailand, pp. 558–563, (2013).

[88] Peel, M.C., Finlayson, B.L., McMahon, T.A., "Updated world map of the Köppen-Geiger climate classification," Hydrology and Earth System Sciences, 11(5), 1633–1644, (2007).

[89] Winslow, J.C., Hunt, E.R., Piper, S.C., "A globally applicable model of daily solar irradiance estimated from air temperature and precipitation data," Ecological Modelling, 143 (3), 227–243, (2001).

[90] Jeffrey, S.J., Carter, J.O., Moodie, K.B., Beswick, A.R., "Using spatial interpolation to construct a comprehensive archive of Australian climate data," Environmental Modelling & Software, 16(4), 309–330, (2001).

[91] Liu, D.L., Scott, B.J., "Estimation of solar radiation in Australia from rainfall and temperature observations," Agricultural and Forest Meteorology, 106(1), 41–59, (2001).

[92] Maddox R.A., McCollum, D.M., Howard, K.W., "Large-Scale Patterns Associated with Severe Summertime Thunderstorms over Central Arizona," Weather and Forecasting, 10(4), 763–778, (1995).

[93] Blanchard, D.O., "A Comparison of Wind Speed and Forest Damage Associated with Tornadoes in Northern Arizona," Weather and Forecasting, 28(2), 408–417, (2013).

[94] Villalva, M.G., Filho, E.R., "Dynamic analysis of the input-controlled buck converter fed by a photovoltaic array," SBA: Controle & Automação Sociedade Brasileira de Automatica, 19(4), 463–474, (2008).

[95] Gao, L., Dougal, R.A., Liu, S., Iotova, A.P., "Parallel-Connected Solar PV System to Address Partial and Rapidly Fluctuating Shadow Conditions," IEEE Transactions on Industrial Electronics, 56(5), 1548–1556, (2009).

[96] Bidram, A., Davoudi, A., Balog, R.S., "Control and Circuit Techniques to Mitigate Partial Shading Effects in Photovoltaic Arrays," IEEE Journal of Photovoltaics, 2(4), 532–546, (2012).

[97] Patel, H., Agarwal, V., "MATLAB-Based Modeling to Study the Effects of Partial Shading on PV Array Characteristics," IEEE Transactions on Energy Conversion, 23(1), 302–310, (2008).

[98] IEC standard voltages, IEC 60038 Edition 6.2, International Electrotechnical Commission (IEC), (1983).

[99] Hine, F., "Fused Salt Electrolysis and Electrothermics," in Electrode Processes and Electrochemical Engineering, New York, NY: Plenum Press, pp. 193–217, (1985).

[100] Mazumder, S.K., "Stability Analysis of Parallel DC-DC Converters," IEEE Transactions on Aerospace and Electronic Systems, 42(1), 50–69, (2006).

[101] Middlebrook, R.D., Cuk, S., "A general unified approach to modeling switching-converter power stages," International Journal of Electronics, 42(6), 521–550, (1977).

[102] Gurvich, L.V., Bergman, G.A., Gorokhov, L.N., Iorish, V.S., Leonidov, V.Ya., Yungman, V. S., "Thermodynamic Properties of Alkali Metal Hydroxides. Part 1. Lithium and Sodium Hydroxides," Journal of Physical and Chemical Reference Data, 25(4), 1211–1276, (1996).

[103] Millero, F.J., Feistel, R., Wright, D.G., McDougall, T.J., "The composition of Standard Seawater and the definition of the Reference-Composition Salinity Scale," Deep-Sea Research Part I: Oceanographic Research Papers, 55(1), 50–72, (2008).

[104] Bureau of Transportation Statistics, U.S. Department of Transportation, Southeast Washington, DC. U.S.A.